HOW NOT TO BE REPLACED
BY A SPREADSHEET THAT TALKS

HOW NOT TO BE REPLACED BY A SPREADSHEET THAT TALKS

GENERATIVE AI FOR THE FUNDAMENTAL INVESTOR

EHSAN EHSANI

NORTHLIGHT EDITIONS
NEW YORK, NY

NORTHLIGHT
EDITIONS

New York, NY

Printed in the United States of America

*To the investors
who seek to master tools
without being mastered by them.*

CONTENTS

PROLOGUE:
A LITTLE FIRE, A LOT OF LIVER

ONCE UPON A MYTH, Prometheus stole fire from the gods and gave it to humanity. The gods, as you can imagine, were not amused. For this act of unauthorized knowledge-sharing, Zeus chained him to a rock where an eagle would show up every day to eat his liver. The liver would regenerate overnight because ancient Greek biology is both cruel and oddly efficient. The whole gruesome cycle would repeat. Forever.

This is, admittedly, a little dramatic. But also not entirely unlike what's happening in Silicon Valley.

Google gave us the fire of transformer models, the architecture behind ChatGPT and pretty much every GenAI application today. Meta built the Llama models and just sort of handed them to the internet like,

"Please don't burn down civilization." OpenAI and Anthropic are doing Prometheus cosplay at industrial scale, unleashing tools that can write code, analyze financial statements, and occasionally invent entirely fictitious statistics with suspicious confidence.

But the funny part is that the fire isn't free. It requires data centers the size of aircraft carriers and enough electricity to make a Bitcoin miner blush. And the companies that brought us this gift are now finding that it's eating their livers. Search revenue is getting cannibalized by AI chatbots. Social media engagement is drifting into generative rabbit holes. Everyone is investing tens of billions into the very thing that may blow up their core business models. I'm not saying it's as bad as being strapped to a boulder while a bird munches on your internal organs for eternity. But it's also not exactly like being gently tickled by clouds.

Anyway. This is a book about how all of this, the fire, the birds, the weirdly expensive GPUs, is changing fundamental investing. It's not a how-to-make-yourself-AI-proof manual. Mostly because AI-proofing doesn't exist, and anyone who promises it probably also sells cybersecurity tips via pop-up ads. What this is about is equipping you with a mindset. One that helps you understand the weird, exciting, occasionally terrifying ways GenAI is reshaping the investing profession. And yes, there are some tips and tricks to help you move faster than the average fund manager still trying to figure out if GPT is a new ETF.

To be a successful investor today, you need two things: the right mindset and the right process. By buying this book, you've already covered the first half. That's 50 percent of the journey. Nice work.

The other half is in my book *Finding Value in Numbers* (Columbia Business School Publishing, forthcoming January 2026, available for pre-order), which covers the process piece. Buy that one too and you'll be well on your way to becoming a billionaire. Or at the very least, someone who understands what a Monte Carlo simulation actually simulates. Plus, each purchase earns me thirty dollars, which might get me and my AI assistant a Shake Shack lunch. She's not a big eater. Mostly just runs on tokens.

So, buckle up. This isn't fire. But it might just light a spark.

See you in Chapter One.

CHAPTER 1
HOW NOT TO BE REPLACED BY A SPREADSHEET THAT TALKS

In finance it is traditional to believe that whatever is happening right now is a once-in-a-generation shift. In 1999 it was the internet. In 2007 it was the iPhone. In 2021 it was meme stocks, which, to be fair, did make some people rich but mostly made some people unemployed in the compliance department. Each time the general idea was that the world had changed forever, and nothing would ever be the same, except that mostly it was the same, just with more screens.

Generative AI is different. This time the change might actually be permanent, and also it might come for you personally. If you have a job that involves reading things and then writing other things about the things you read, this is awkward news.

The sell side has already been through a smaller version of this. In the late 1990s, if you worked at a big U.S. investment bank you could probably name five people who covered regional airlines, and at least one of them had an intern who specialized in "small-to-medium-sized West Coast routes." Entire beverage teams debated whether Monster Energy belonged in "non-alcoholic beverages" or "other consumer staples," and they were both right in their own way.

Then headcount started falling. Partly because of technology. Partly because of regulation. Partly because management discovered the joy of "synergies," which in this context means you cover twice as many stocks for the same pay. Reg FD and MiFID II did not help. Trading commissions collapsed and research became a cost center, which is a phrase that in corporate finance is generally a prelude to headcount being described in the past tense.

The numbers are not subtle. Twenty years ago, a single bulge-bracket bank might have had 300 equity research analysts. By the early 2020s, the top ten combined had fewer than that.

Generative AI takes that trend and speeds it up in the way an avalanche speeds up after it starts moving. You could imagine a future where a team that used to be ten people is now three, and the official line is that the output is "enhanced" because one of those three has a laptop with a chatbot.

Now imagine a fund manager with three analysts. He has three analysts. They are all MBAs, which in finance is code for "expensive." Each one earns $150,000 in base salary plus a bonus for doing things like building models and summarizing filings. Then one day the manager stumbles into a generative AI tool, spends a weekend experimenting with it, and realizes that if you combine it with some finance-specific models and $50,000 worth of decent data feeds, it can do a solid imitation of one of those analysts.

That analyst is Oskar. Oskar is good at pulling numbers, writing summaries and explaining them in bullet points. So is the AI. The AI works nights, does not ask for a bonus and has no views on whether Monster Energy should be valued like a soda or like a lifestyle brand. The fund manager keeps two analysts and lets Oskar "pursue other opportunities," which is HR language for "good luck out there."

This is the Oskar problem. It is the possibility that someone realizes they can get most of your output for a fraction of your cost and decides that "most" is good enough.

In the past the automation risk in finance was mainly about the typing part of the job. You could lose your seat if Excel macros could do your formatting, but the actual thinking part was safe. This is different. Generative AI produces work product that looks like thinking. Sometimes it even is thinking, or something that is close enough to pass in a meeting.

The uncomfortable truth is that you do not have to be replaced by the best AI. You only have to be replaced by the AI that your boss thinks is fine.

Avoiding that outcome is not about being irreplaceable. Nobody is. It is about making yourself more valuable with the tools than without them. If you can make the AI your assistant, your data retriever, your first-draft generator and your tireless junior associate, then you spend less time getting to the starting line and more time deciding where to run.

This book will not make you Oskar-proof. Nothing will. But it will give you a running start at using generative AI in a way that makes you the person with the AI, not the person replaced by it.

Oskar, for what it's worth, is fine. He works at a fintech now. He does presentations about the blockchain.

You are here because you would prefer not to be Oskar.

CHAPTER 2
THE PROMPT IS THE JOB

IF GENERATIVE AI IS A MAGIC MACHINE that gives you answers, then the "prompt" is the magic spell. And like most magic spells, the difference between something powerful and something embarrassing is all in the wording. Say the spell correctly, and you conjure something impressive. Mumble it vaguely, and you end up with a toad, or, in AI terms, a generic Wikipedia paragraph about "the semiconductor industry" with no actual insight.

A "prompt" is just what you tell the AI before you ask it to do something. It is the setup, the context, the stage directions. We call it a prompt because "thing you tell the robot so it does the right thing" was too long. In finance, this is familiar: a pitch memo, a stock screen, a model. They are all just prompts in disguise.

But here's the thing: with these AI models, prompt hygiene matters. And if you are in the business of turning information into decisions, your ability to construct a good prompt might be the single biggest short-term edge you have. That edge may not last forever. Eventually, everyone will be good at this. For now, though, you can get paid for being better at asking the question.

Here is the dirty secret. If you ask the same basic thing everyone else is asking, you will get the same basic output everyone else is getting. And that means no edge.

So, you need a structure. You need your prompts to be C.R.E.A.T.I.V.E.F.

C.R.E.A.T.I.V.E.F. PROMPT FRAMEWORK

C – Context: Give the model enough background so it knows the playing field. "Nvidia, Q2 2025 earnings call" is context. So is "You are acting as a buy-side analyst." Without context, the AI is like a new intern who just walked in from lunch and missed the morning meeting.

R – Role: Tell the AI who it is in this scenario. "You are a sell-side analyst" will yield different results than "You are a financial journalist" or "You are a skeptical hedge fund PM." Models are people-pleasers, so if you tell them who they are, they will play the part.

E – Expectation: Spell out what "good" looks like. Is it concise? Highly detailed? Optimized for speed?

A – Audience: Who is this for? Your boss? An investment committee? The client who reads only bullet points? The AI will write differently for each.

T – Task: Be clear about what the AI is doing. "Summarize" is not the same as "summarize and compare to prior quarter." The more specific you are, the fewer revisions you will need.

I – Input: Give the AI the data or direct it to find the data. Do not just say "analyze NVDA earnings." Provide the transcript or tell it exactly where to look.

V – Voice: Tell the AI the tone you want, whether formal, conversational, skeptical, or optimistic. Otherwise, it defaults to "bland corporate newsletter."

E – Edges (boundaries): Limit the scope. Tell it what not to do. ("Do not speculate beyond management's statements.") AI loves to be helpful by making things up. Boundaries keep it honest.

F – Format + Follow-up flow: Specify output structure and what the next step will be. If you want bullet points under H3 headers, say so. If you will ask for a valuation model after the summary, set that up in the prompt.

A BAD PROMPT

Here is what not to write: "Analyze Nvidia's Q2 2025 earnings call." That's it. No context, no role, no guidance, no format. This is the equivalent of walking into a meeting and saying "Talk about the company" without telling anyone which company or why. You will get something vague, incomplete, and probably wrong in interesting but useless ways.

A GOOD PROMPT

Now let's take your "Earnings Call Analysis Annotated Best Practice Template" and turn it into a fully CREATIVEF-compliant prompt. But before that, let's talk about the components you need. When you think of prompts, it helps to think about securities filings. In theory, you could get by with the short form. A few lines, the bare minimum, just enough to get the job done. That is what **CREATIVEF** gives you: the skeleton of a good prompt. Context, Role, Expectation, Audience, Task, Input, Voice, Edges, Format. It is like the summary at the top of a prospectus. It tells the AI who it is supposed to be, what it is supposed to do, who it is writing for, and how the output should look. With CREATIVEF alone, you are already "compliant," already producing something that works.

But of course, investors rarely rely only on the summary. They read or at least, they demand the footnotes, the exhibits, the appendices. In prompt design, that extra layer is the **Analyst Workflow**. It is not technically necessary, but it fills in the gaps, spells out the mechanics, and removes any ambiguity. It says: here are the sources you should consult, here are the key numbers to extract, here is the structure you must follow, here is how you should format every line. The Analyst Workflow is the boring part, the fine print, the bit that nobody admires but everyone depends on.

So the real choice is not whether to use CREATIVEF, you always should. The choice is whether to stop there or to add the Analyst Workflow too. If you are writing a one-off prompt for yourself, the framework is usually enough. If you are building something you want to reuse, share, or institutionalize, then adding the workflow is what makes the prompt bulletproof. It is the difference between a decent sketch and a polished playbook, between "works once" and "works every time."

THE C.R.E.A.T.I.V.E.F. COMPONENT

Here is the CREATIVEF component of a good prompt in the context of Nvidia's Q2 2025 earnings call:

Context: You are assisting in the analysis of Nvidia Corporation's Q2 2025 earnings call for internal buy-side investment research.

Role: You are a buy-side investment analyst preparing a concise, decision-useful summary for a portfolio manager.

Expectation: Produce a thorough, well-structured Markdown summary of the earnings call that highlights key numbers, qualitative insights, and potential investment implications. The output should be clear, accurate, and immediately usable for portfolio decision-making.

Audience: Senior portfolio manager with deep sector knowledge but limited time.

Task: Summarize and analyze the earnings call transcript, including management's prepared remarks and

Q&A. Identify headline KPIs, secondary metrics, guidance changes, notable strategic initiatives, and sentiment drivers from Q&A.

Input: Use the official Nvidia Q2 2025 earnings call transcript from SEC 8-K, Nvidia IR website, Refinitiv, FactSet, or Seeking Alpha.

Voice: Professional, precise, and analytical, with occasional emphasis on items of strategic or valuation significance.

Edges: Do not include speculative information not stated by management. Avoid vague commentary such as "management was optimistic" without citing supporting remarks.

Format + Follow-up flow: Output should use H2 (##) for major sections (e.g., Key Metrics, Guidance, Q&A) and H3 (###) for subsections. Use bullet points for all items under each heading. For numeric data, include the actual value, % change, and comparison period where possible. Follow this summary with a 3–5 sentence "next steps" section suggesting follow-up research or analysis.

ANALYST WORKFLOW COMPONENT (FOR THE AI)

And here is the Analyst Workflow component of the prompt for Nvidia's Q2 2025 earnings call.

SOURCE TRANSCRIPT

- SEC 8-K
- Company IR website
- Refinitiv, FactSet, Seeking Alpha

EXTRACT AND ANALYZE

- Headline KPIs (revenue, EPS, margins, etc.)
- Notable secondary metrics
- Guidance updates
- Strategic initiatives and forward-looking commentary
- Q&A sentiment movers

FORMATTING RULES

- H2 for main sections, H3 for sub-sections
- Bullet points for all data and commentary
- Use % and $ symbols appropriately
- If no material info, say "No material remarks"

OUTPUT STRUCTURE

- Management attendees

- Key metrics

- Other metrics

- Guidance

- Future plans

- Q&A highlights

Follow-up: Indicate readiness for valuation modeling prompt once summary is complete.

Ending the prompt: I will now give you the company name, ticker, and fiscal period: **Nvidia (NVDA), Q2 2025**

EXAMPLE OUTPUT (NVDA Q2 2025)

This is where theory meets reality. You have heard me go on about context, roles, boundaries, and all the rest of the CREATIVEF alphabet soup. Now here is what it actually looks like when you feed that kind of well-structured prompt to the machine and point it at something real, in this case Nvidia's Q2 2025 earnings call. Notice a few things as you read it: you could drop this straight into an investment committee memo without changing a word, the headings and bullet points make it skimmable for anyone who only has three minutes before their next Zoom call, and it is precise enough to be trusted while structured enough to be automated into models or decks.

Now to the output: If you run the Nvidia Q2 2025 prompt through an AI, the result looks like an analyst memo. In fact, it looks like a very long analyst memo, the kind of thing that would make perfect sense in an investment committee binder but might be a little heavy for a book about prompt design. And that makes sense: the framework is supposed to produce something comprehensive, precise, and decision-useful, not just clever words on a page. For an analyst who needs to make a real portfolio decision, the length is a feature, not a bug. For a reader who is just trying to make it through Chapter 2 without falling asleep, it is probably less helpful. So what follows is a sampler of the output, trimmed for readability, but still carrying the flavor of what a CREATIVEF-compliant output looks like in practice.

MANAGEMENT ATTENDEES

- Jensen Huang, President & CEO

- Colette Kress, EVP & CFO

- Stewart Stecker, Investor Relations

KEY METRICS (Q2 FY2025)

- Revenue: $30.0B (+15% q/q; +122% y/y)

- GAAP EPS: $0.67 (+168% y/y)

- Gross margin: 75% (down q/q, up y/y)

SEGMENT HIGHLIGHTS

- Data Center: $26.3B (+154% y/y), still the growth engine

- Gaming: $2.9B (+16% y/y)

GUIDANCE (Q3 FY2025)

- Revenue: $32.5B ±2%

- Margins: mid-70s%

FUTURE PLANS

- Blackwell ramp starting Q4; Hopper demand still strong

- Networking (Ethernet for AI) adoption accelerating

Q&A HIGHLIGHTS

- Inference about 40% of DC revenue, seen as durable

- China revenue recovering but below pre-controls

INVESTMENT IMPLICATIONS

- Strong top-line momentum, diversified revenue drivers

- Blackwell ramp gives visibility into next node

- Margins to watch during transition

WHY THIS WORKS

This is not just about being fancy. The structure makes the AI more accurate and efficient. Telling it the role changes the style. Giving it edges prevents hallucinations. Specifying format makes the output plug-and-play for whatever your next step is: a valuation model, a client email, or an investment committee memo.

And yes, Markdown formatting matters. In Markdown, ## makesan H2 heading, and ### makes an H3 heading. Keeping your structure consistent means you can automate the movement of AI output into reports, dashboards, oreven slide decks without hand-editing.

In other words, structure is not window dressing, it is leverage. And once you see it that way, every prompt becomes less of an experiment and more of a repeatable tool you can actually rely on.

META NOTE FOR THE READER

You will notice that the prompts in this chapter are like a five-course tasting menu, elaborate, layered, and full of nuance. Later in the book, the prompts get more streamlined. This is not because we suddenly forgot how to prompt well. It is because if we wrote every single prompt in full CREATIVEF glory, this book would weigh more than a Bloomberg Terminal and cost as much to print. Consider this chapter your prompt masterclass, and later chapters the quick-reference guide for when you are on the move or cornered at a cocktail party by someone asking for your best idea.

PARTING WORDS ON PROMPT HYGIENE

Prompt hygiene is like personal hygiene. When it is good, no one notices. When it is bad, everyone notices. And the advantage you have now is that most people still have bad prompt hygiene. They throw vague, lazy instructions at the AI and then complain about the answers.

For now, the people who write good prompts get better results. Eventually, everyone will figure this out. But in the meantime, your edge might just be knowing how to talk to the machine better than the analyst in the next seat.

Your career is not going to be replaced by AI. It is going to be replaced by the person sitting next to you who knows how to tell AI exactly what to do. And that person might as well be you.

CHAPTER 3
HOW TO TALK TO YOUR ROBOT WITHOUT MAKING IT WEIRD

HERE'S THE THING ABOUT GENERATIVE AI for investing: it is like having a super-smart intern who has read every finance book, every earnings call transcript, and also 80% of the internet… but who is also a compulsive people-pleaser.

This intern is one of those who desperately want you to like them. If you say, "Hey, tell me why Tesla is the greatest company in the world," they will nod, smile, and start telling you why Tesla is the greatest company in the world. They will not, unless specifically instructed, say, "Well actually, here are 14 reasons it might crash and burn." That is not because the AI is secretly short Tesla. It is because you told it what you wanted to hear. It is just giving you what you asked for.

Generative AI can help you generate investment ideas like a beast, a caffeinated never-sleeping portfolio-brainstorming beast. But my friend, if you do not talk to it in the right language, it will happily give you ideas that are biased, outdated, and possibly the intellectual equivalent of a "hot penny stock tips" email from 2004.

Before you let the machine spit out your next great portfolio move, let us talk about three big cautions in AI-driven idea generation. These are not theoretical. They are the kind of things that, if ignored, will make your beautiful AI-assisted investment process look like a flaming dumpster rolling downhill into your capital allocation strategy.

"YOUR STOCK PICKS ARE OK-- BUT I MISS THE WARM PERSONAL TOUCH OF A HUMAN BROKER."

CAUTION 1: AVOIDING CONFIRMATION BIAS

AI is like a mirror that reflects your own biases back at you, but with citations and bullet points. If you walk up to it and say:

"Explain why Nvidia is undervalued and will continue to outperform the market in the next 12 months."

It will do exactly that. It will Google in its brain, assemble all the bullish analyst notes, and enthusiastically tell you why CEO Jensen Huang is basically Steve Jobs plus Batman. It will not stop and say, "Hey, maybe you should think about valuation risk or competitive threats from AMD."

The problem is that you have framed the question in a way that assumes the conclusion. This is the fastest way to turn AI into a confirmation-bias amplifier. You will feel extra smart, and extra wrong if the trade blows up.

Bad Prompt Example: "Explain why Nvidia (NVDA) is undervalued and will continue to outperform the market in the next 12 months."

Good Prompt Example: "Analyze Nvidia (NVDA) as a potential investment, presenting both bullish and bearish cases, key risks, catalysts, and valuation metrics. Include recent financial performance and competitive landscape.

Why this works: The good version forces the AI to gather both sides of the story. You are essentially asking it to play devil's advocate against itself. It is no longer just your bias machine. It is your balance machine.

CAUTION 2: VERIFYING SOURCES AND NUMBERS

If you have been around finance long enough, you know that nothing is more dangerous than bad numbers that sound authoritative. In the AI world, this is called "hallucination," which is a polite way of saying "making stuff up."

If you ask AI, "What is Nvidia's current P/E ratio and quarterly revenue growth?"…it might tell you the right answer. Or it might give you numbers from two years ago, or from a blog post that miscalculated them, or from a half-remembered Yahoo Finance scrape. In other words, garbage in, garbage out, but dressed up in a suit and tie.

Bad Prompt Example: "Tell me Nvidia's current P/E ratio and quarterly revenue growth."

Good Prompt Example: "Find Nvidia's (NVDA) most recent P/E ratio, market cap, and last quarter's YoY revenue growth from reliable sources such as the company's 10-Q, Bloomberg, or Yahoo Finance. Provide the date of the data and link to the source."

Why this works: The good version makes the AI cite sources and timestamps. Now you know whether that P/E is from last quarter or the day Lehman collapsed.

CAUTION 3: BALANCING NOVELTY WITH PROVEN CRITERIA

We all want to find the next big thing. But novelty for novelty's sake is a siren song that has led many an investor to the rocky shores of zero.

If you ask AI: "List 10 small-cap AI companies no one is talking about that could be the next Nvidia."…you will get a list. Oh, you will get a list. It will be filled with cool names you have never heard of, and you will get a little dopamine hit just reading them. But it will not necessarily tell you that these companies are hemorrhaging cash, or that their core product is a chatbot for ferrets. Novelty bias is real. And AI, unless told otherwise, assumes novelty is what you want.

Bad Prompt Example: "List 10 small-cap AI companies no one is talking about that could be the next Nvidia."

Good Prompt Example: "Identify 10 lesser-known AI-related companies (market cap $500M–$5B) with at least 20% YoY revenue growth, positive free cash flow, and P/E < 25. Include sector classification, recent news, and fundamental metrics."

Why this works: You still get the novelty, but it is filtered through actual, testable financial criteria. Think of it as novelty with guardrails.

THE BEST PRACTICE PROMPT STRUCTURE

Here is a reusable template that bakes in all three cautions: avoiding bias, verifying sources, and balancing novelty with proven investment criteria. You can plug in any company, sector, or theme and have a prompt that gives you an intelligent, balanced, and data-backed output.

Template: "I want to explore potential investments in **[company/sector/theme]**. Please:

- Present both bullish and bearish arguments, including key risks and catalysts.

- Use the latest data from reliable sources (e.g., 10-K/10-Q, Bloomberg, FactSet, Yahoo Finance) and cite the date and source.

- Apply these objective screening criteria: [insert valuation, growth, or quality metrics].

- Include both established leaders and emerging players, noting how each aligns with proven investment principles.

- Flag any data limitations or uncertainties so I know where further manual verification is needed.

- Present both bullish and bearish arguments for each company.

- Use the latest available data (EPS, P/E, market cap, YoY revenue growth) from verifiable sources like Bloomberg or the company's filings, citing date and source.

- Filter for companies with market cap > $10B, P/E < 30, and positive free cash flow.

- Include both large caps (e.g., Nvidia, TSMC, AMD) and promising mid-caps, explaining how each meets or does not meet established investment criteria.

- Flag any missing or uncertain data points."

PARTING TIPS FOR NOT MAKING YOUR AI SAD AND YOUR PORTFOLIO SADDER

- Always ask for the other side of the story. Even if you think you know the answer, make the AI argue against it.

- Timestamp your data. If you would not trust a chart without an axis label, do not trust AI data without a date.

- Be specific about filters. Market cap ranges, P/E thresholds, FCF positivity, whatever matters to you, spell it out.

- Avoid "hot take" prompts. They are fun for entertainment but terrible for capital allocation.

- Think of AI as an analyst who is always on probation. It can do great work, but you double-check everything before making a move.

The whole point of using AI in idea generation is leverage. You are not outsourcing your thinking. You are supercharging it. But without good prompt hygiene, you are basically supercharging your mistakes.

A bad prompt is like handing your intern a blank check and saying, "Just buy whatever is exciting today." A good prompt is like handing them a detailed investment memo template and saying, "Fill this out with facts, not feelings, and we will talk."

Basically, you need to remember that generative AI will not make you a better investor by itself. It is not a magic crystal ball. It is more like a magic parrot. It will repeat back whatever bias, assumption, or fantasy you whisper into its ear, just louder and with more bullet points. Your job is to make sure you are whispering the right things.

CHAPTER 4
IDANM
(AND YES, IT'S ANOTHER ACRONYM)

AT THIS POINT IN THE BOOK YOU might be wondering, "Okay, I understand that I should avoid becoming Oskar, and I understand how to talk to the machine without embarrassing myself, but how does all this fit into the actual job of investing?"

Fair question. We need a conceptual framework. And yes, I realize that in finance the phrase "conceptual framework" usually means "something we made up to justify the chart on page six." But hear me out.

A framework is helpful here because it gives us a map. Without a map, you risk getting lost in the infinite things generative AI could do instead of focusing on the few things it should do for you as a fundamental investor. It is like running a stock screen without any filters. You get back every company in the world and spend the rest of your career in Excel purgatory. Also, a framework gives us an excuse to make another acronym. Which we are going to do.

YES, ANOTHER ACRONYM

We just finished C.R.E.A.T.I.V.E.F. for prompts and now we are doing I.D.A.N.M. No, you do not have to memorize them. No, there will not be a test. The point is not to turn your brain into a file cabinet for letters. The point is to have a convenient hook for remembering the big buckets of where AI fits in your workflow. IDANM stands for:

- **Idea Generation** – Finding Stuff Worth Caring About

- **Data Gathering & Processing** – Collecting the Haystack and the Needle

- **Analysis & Valuation** – Turning Numbers into Opinions

- **Narrative & Thesis Development** – Making the PowerPoint That Gets You Paid

- **Monitoring & Risk Management** – Watching the Stuff You Bought Like a Hawk

That is it. That is the whole map. Every part of your job as a fundamental investor can be stuffed into one of these buckets, and every one of these buckets can be made better, faster, or less miserable with generative AI.

IDEA GENERATION

Every good investment starts with an idea, which sounds profound until you realize that most of your "ideas" come from reading about other people's ideas. The role of AI here is not to magically dream up a perfect trade that nobody else in the world has thought of, although if it does, you are welcome. It is to expand your surface area for idea discovery.

Instead of reading every 13F, you can have AI summarize them and flag the unusual trades. Instead of scanning headlines all day, you can have it identify companies hitting certain metrics or trends. It is not that you stop thinking. It is that you stop pretending you have time to manually read 300 news articles a day.

DATA GATHERING & PROCESSING

Most of the job of investing is not having the idea but proving the idea is not stupid. This means finding data, cleaning it, organizing it, and turning it into something usable. This is the haystack stage and the needle stage.

AI is exceptionally good at both. It can scrape and summarize hundreds of pages of filings in minutes. It can merge datasets, normalize terms, and flag anomalies before you even get to Excel. It does not get bored and it does not get carpal tunnel. And if you write your prompts well, see C.R.E.A.T.I.V.E.F., it will give you exactly what you need in a usable format the first time.

ANALYSIS & VALUATION

At some point you have to decide what something is worth, and unfortunately AI still cannot see the future. But it can help you get there faster.

It can build and stress-test valuation models based on your inputs. It can run scenario analyses on different margin assumptions or FX rates. It can explain why a DCF is giving you nonsense numbers, which might be the first time anyone has ever wanted to explain that to you. And it can do this without asking for an end-of-year bonus.

NARRATIVE & THESIS DEVELOPMENT

Finance is storytelling with numbers. The narrative is what makes your analysis persuasive to a portfolio manager, an investment committee, or a client. AI can help you take your spreadsheet full of conclusions and turn it into something that reads like a coherent argument.

This is where tone and audience matter. A thesis for an IC deck should not sound like a blog post, and a client letter should not sound like an IC deck. AI can adapt your message to the right voice and format instantly. It can draft, rewrite, and polish until the story is tight.

Monitoring & Risk Management

Once you own something, you have to watch it. Closely. This is where AI shines in keeping you up to date without you spending your life glued to a terminal.

You can set up AI agents to monitor news, filings, sector developments, even changes in tone from management on earnings calls. It can flag risk events in near real-time. It can summarize quarterly developments into exactly the kind of update you wish you had time to write for every holding.

"THE COMPUTER IS CLAIMING ITS INTELLIGENCE IS REAL, AND OURS IS ARTIFICIAL."

HOW WE WILL USE IDANM IN THIS BOOK

Each of these five areas will get its own deep-dive chapter later. That way, you get both the big picture and the "here is what to actually do tomorrow morning" detail. We will start with the areas where GenAI is already making a clear difference, Idea Generation and Data Gathering, and move toward the areas where it is still emerging but worth experimenting with, Narrative and Monitoring. And yes, we will keep making fun of acronyms along the way.

Without a structure like IDANM, AI adoption can feel like a random set of tricks. One day you are summarizing an earnings call, the next you are asking it to draft an email to compliance, and then you forget to use it for a week. With the structure, every part of your process has a "slot" where AI might fit.

The risk of not having this structure is the same risk you have without an investment framework. You either underuse AI or use it in inconsistent, low-value ways. Worse, you will spend time on things that do not move the needle.

PARTING WORDS

Think of IDANM as the cover page of the map. CREATIVEF tells you how to talk to the machine, IDANM tells you where in your workflow the machine belongs. You need both. Without CREATIVEF, you are asking the wrong questions. Without IDANM, you are asking the right questions in the wrong places.

Generative AI will not magically make you a better investor. But it will happily make a better investor out of someone else, and then take your seat, if you ignore where and how to use it.

CHAPTER 5
FROM CNBC TO GPT:
LEVELING UP YOUR IDEA PIPELINE

Generative AI can come up with investment ideas for you. Let that sink in for a second. You can be sitting in your chair, sipping your third coffee of the morning, and instead of scrolling through your usual list of newsletters, 13F filings, and Twitter arguments, you can just type a sentence into your laptop and watch a shortlist of potential investments appear. It feels like magic, except it is just math pretending to be magic.

For decades, fundamental investors have worked the same basic idea pipeline. You might start your day reading a Substack letter from a well-known analyst who writes in all caps when he is excited. Or maybe you catch pundit Charlie on CNBC pounding the table for a mid-cap retailer whose name he mispronounces. You might hear about a new idea at a dinner with other value investors where the steak is overcooked but the stock pitch is medium rare. Sometimes you stumble on it in a hedge fund letter, read about it in a market report buried in footnote 17, or from a big 13F filing that hints a respected investor just bought 4% of a little-known industrial company in Ohio.

Then there are the screens. You fire up Bloomberg or Capital IQ and run your favorite filters for low EV/EBITDA, high free cash flow yield, or insider buying. You read sell-side initiation reports, thumb through trade journals, attend conferences, and talk to people who actually operate in the industry. You might even, in a fit of masochism, sit through a full hour of an earnings call where the CFO insists that "headwinds are transitory" for the fourth quarter in a row.

That is the traditional game. It works. But it is mostly reactive. You are constantly waiting for something to land in your lap. A headline. A conference invitation. A letter from a manager who happens to like the same style of investment you do.

AI MAKES YOU THE IDEA HUNTER, NOT THE IDEA HUNTED

Generative AI changes the dynamic. Instead of reacting to what crosses your desk, you can go out and interrogate the entire ocean of data to surface opportunities. You can ask questions the same way you would to a junior analyst, but with the ability to synthesize not just a few reports but potentially every earnings transcript, blog post, and patent filing on the topic.

You are no longer just reading the news; you are actively creating it in your own head. AI lets you go from "I wonder if EV adoption is creating any interesting battery supply chain plays in Southeast Asia" to a ranked list of companies, their recent financial performance, relevant news catalysts, and comparable names in other regions in minutes.

Where the old approach was "hope someone brings me something interesting," the AI-enhanced approach is "I am going to map this opportunity space myself and see where the overlooked pockets are."

THEMATIC EXPLORATION WITH AI

One of the most exciting uses of generative AI is thematic exploration. Traditionally, identifying an emerging theme meant reading a lot of scattered sources and slowly piecing the story together. You might catch an early hint from a regulatory filing in the EU, pair it with a press release from a Chinese supplier, and then confirm your hunch when you see a small-cap in Canada quietly doubling its capex.

Now you can tell AI: "Identify sectors and companies that could benefit from the EU's new carbon border adjustment mechanism, focusing on manufacturing and energy efficiency plays."

It can digest hundreds of news articles, government releases, and analyst notes, then present you with a list of companies organized by potential exposure. It can spot trends you might not have had the bandwidth to connect, like how a regulatory change in the US trucking industry is sparking demand for certain telematics providers.

AI can also pull from sources you might not think to check. That could mean obscure blog posts, niche industry reports, or even conference call Q&A sessions where a CEO casually mentions expanding into a market that later becomes a huge revenue driver. It is like having a team of thematic scouts working across multiple continents without you booking a single flight.

COMPANY SCREENING THAT GOES BEYOND NUMBERS

You know the standard drill: run a screen for certain valuation metrics, then eyeball the results to see what is interesting. It works, but it has limits because it only captures what you explicitly filter for. AI lets you blend the quantitative with the qualitative.

You could say:

"Screen for US-listed companies with market cap above $500M, P/E under 15, and positive free cash flow that have mentioned 'vertical farming' in their last two earnings calls."

That is a hybrid filter, half traditional metrics, half unstructured text search. You might find that a logistics company you never considered is actually building a meaningful side business supplying vertical farm operators.

You can also rank candidates based on narrative fit with your thesis, something traditional screeners cannot do. For example:

"From the Russell 2000, identify the top 15 companies most aligned with the thesis that US reshoring of manufacturing will accelerate in the next five years. Rank by relevance to reshoring and recent capex activity."

CONNECTING THE DOTS ACROSS SECTORS AND MARKETS

One of AI's more underrated skills is linking ideas across seemingly unrelated areas. Say you read that Toyota is investing heavily in solid-state batteries. A traditional investor might stop at "good for Toyota, bullish for certain battery makers." An AI-enhanced investor could prompt:

"Identify public companies outside Japan that could benefit indirectly from Toyota's solid-state battery program, focusing on materials suppliers, testing equipment manufacturers, and logistics providers."

Suddenly you are looking at Australian lithium miners, German lab equipment makers, and a small US port operator that just expanded its capacity for hazardous material handling. That kind of cross-market mapping used to take weeks of networking and research. Now it can be a single query.

AI can also connect supply chain dots that are not obvious. For example, if Nvidia announces a new data center chip, you could ask:

"Map the supply chain for Nvidia's new H200 GPU, identifying public companies in packaging, testing, and distribution that could see revenue impact."

You might find a small Singapore-listed testing company that is a pure play on this growth but not yet on analysts' radar.

COMPETITIVE LANDSCAPE ANALYSIS AS AN IDEA SOURCE

Competitive mapping has always been part of fundamental research. But AI can make it a near-real-time process.

Suppose you are looking at MercadoLibre in Latin America. You could prompt:

"Identify emerging competitors to MercadoLibre in Brazil and Argentina, including any notable venture-backed startups that have recently raised funding."

You could then take it further:

"From that list, identify any public companies with indirect exposure to those startups through partnerships, infrastructure contracts, or payment processing."

It can also surface early indicators of competitive shifts. For example:

"Track US-listed retail companies that have increased logistics-related job postings by more than 30% in the last six months, and identify whether this aligns with market share expansion plans."

A FEW WORKFLOW EXAMPLES

Here is a simple example of a thematic prompt that moves from broad to narrow:

- **Step 1:** "Generate a list of underfollowed public companies in Southeast Asia benefiting from EV battery adoption."

- **Step 2:** "From that list, filter for companies with positive free cash flow, YoY revenue growth above 15%, and market cap under $3B."

- **Step 3:** "Provide a summary of each company's main growth drivers, key risks, and most recent quarterly results."

You can use the same iterative narrowing for other themes. For instance, you might start with an initial query such as "List public companies globally that could benefit from the rise in generative AI tools for enterprise use." From there, you refine it into something more specific, like "Filter for companies with P/E below 25, market cap under $10B, and over 30% revenue from enterprise SaaS." Finally, you bring it down to an actionable level with a query such as "Provide 3-year financial trends and summarize major recent partnerships for each company."

WHY AI-ENHANCED IDEA GENERATION WORKS

The power is in speed and scope. Instead of spending days or weeks gathering scattered bits of information, you can assemble a coherent picture in hours. You can move from macro theme to micro target with far less friction.

It also lets you explore areas you would not normally touch because the cost of curiosity is lower. In the old model, exploring a niche like Chilean copper smelting meant digging through trade journals, paying for industry reports, and hoping you did not waste your time. Now you can do a quick AI sweep to see if it is worth a deeper dive.

PARTING TIPS BEFORE YOU START ASKING THE MACHINE EVERYTHING

- Start broad, then go narrow. The first query should open the map, not hand you the final destination.

- Mix quant and qual. Use both financial metrics and narrative factors in your filters.

- Cross-check with human sources. AI should surface ideas, not replace due diligence.

- Look for indirect plays. Supply chain and ecosystem opportunities are often where the underpriced assets live.

- Iterate. The best lists come from refining your prompts two or three times.

The old way of sourcing ideas was like fishing in a pond you knew well, hoping for a good catch. Using generative AI is like suddenly being handed a map of every pond, river, and ocean on earth, along with a list of what is biting where. Just remember, the map is not the fish. You still have to reel it in yourself.

CHAPTER 6
USING AI TO TAME
THE RESEARCH MONSOON

There is no end to reading this stuff. That is the first thing you learn when you start doing serious investment research. There are more filings, transcripts, industry reports, and news articles than any single person could ever consume. It's like trying to catalog raindrops in a monsoon.

If you are Warren Buffett, you might actually enjoy this. Buffett is famous for reading hundreds of pages a day, from annual reports to trade magazines, with the same enthusiasm most people reserve for dessert menus. He can skim a 10-K, extract the important bits, and still have time to go get a cheeseburger and a Cherry Coke before the rest of us finish the risk factors section. For him, it's less a chore than a lifelong habit, like breathing. The rest of us, meanwhile, are still trying to remember where we left our highlighter.

But you are not Warren Buffett. And you probably have other things to do besides reading footnotes about inventory accounting methods for the next eight hours. Even if you love the work, there is a point where the sheer volume becomes the enemy. You cannot research every angle on every company, so you triage. That means you miss things.

THE PAIN POINTS OF TRADITIONAL RESEARCH

Let's be honest. Manual review of filings, transcripts, and reports is the investment equivalent of mowing a football field with nail scissors. You can do it, but you will hate yourself halfway through.

The first problem is volume. A single large-cap company might produce thousands of pages of filings, investor presentations, and press releases in a year. Add sell-side research, industry publications, regulatory announcements, and you are buried.

The second problem is fragmentation. The data you want is scattered across paywalled databases, public records, PDF downloads, and occasionally some journalist's tweet. You spend as much time finding the information as you do analyzing it.

Then there is note-taking and cross-referencing. Every analyst has a system, whether it is a perfectly organized Evernote library or a chaotic mess of Word docs, Excel sheets, and Post-it notes. Either way, it takes time to collect the facts, sort them, and connect them across sources. And that is before you even start comparing this quarter's results with last year's or reading between the lines of management's language.

HOW GENERATIVE AI CHANGES THE GAME

Generative AI does not replace the need for research, but it changes the economics of it. What used to take a day can now take minutes. It can read faster than you, remember more than you, and never loses its place in a footnote.

Instead of manually slogging through a 250-page 10-K, you can feed it into AI and say:

"Summarize the 10-K for Deere & Co (DE), highlighting revenue by segment, recent strategic initiatives, and any changes in risk factor language compared to last year."

In under a minute you get a clean digest. You can still read the original if you want, but you now have a roadmap of where to focus your time.

For an earnings call, you could prompt:

"Summarize Tesla's Q2 2025 earnings call transcript. Include major announcements, guidance changes, analyst questions about production capacity, and any tone shifts from the previous quarter."

The AI will do the heavy lifting. You still apply your judgment, but the work of wading through 90 minutes of transcript is gone.

"THE ALGORITHMS MANAGING YOUR FUNDS SAY 'BUY AI STOCKS'... "

EXTRACTING WHAT MATTERS AND COMPARING OVER TIME

One of the most powerful tricks is asking AI to highlight changes between filings. For example:

"Compare Apple's 2024 and 2023 10-K risk factors. List what was added, removed, or reworded."

This is pure gold for catching subtle shifts in disclosure that might signal a brewing issue or a new growth area.

You can go further and have it extract structured data from qualitative sources. If a company buries its segment revenue deep in the MD&A section, AI can turn it into a neat table for you. You can even normalize terminology across peers. For instance, "net interest margin" might be called different things in different bank filings; AI can standardize them so you can compare across the sector without confusion.

TURNING UNSTRUCTURED INTO STRUCTURED

Traditional research involves a lot of manual structuring. You might pull revenue by geography from three different 10-Qs, paste it into Excel, and then try to line up the definitions. AI can do that automatically: "From Caterpillar's last three annual reports, extract revenue by segment and geography into a table. Use consistent naming for segments across years."

Once you have structured data, comparison becomes instant. You can see trends without manually digging through PDFs.

You can also tag content automatically. If you are tracking inflation references across your portfolio, you could ask: "Tag all mentions of inflation in the Q2 2025 transcripts for Dow 30 companies, noting sentiment (positive, negative, neutral)." Now you have a thematic map without having to read every line yourself.

INTEGRATING MULTIPLE SOURCES

The real magic is when you combine multiple streams into one AI-readable workspace. Imagine pulling in SEC filings, broker research, alternative data, and news coverage, then asking: "Summarize the investment case for Union Pacific (UNP) based on the last two quarters of filings, major news, and analyst commentary. Highlight any contradictions between management statements and analyst expectations."

AI can spot when management says "on track" while analysts quietly lower earnings forecasts. It can detect inconsistencies that would otherwise require hours of careful reading.

You can also use it to create dynamic knowledge graphs. For example: "Build a map linking all companies mentioned in connection with hydrogen fuel cell development over the past 12 months, including suppliers, customers, and joint ventures." This kind of cross-entity mapping used to require a full research team. Now it can be automated.

REAL-TIME COMPETITIVE AND INDUSTRY MONITORING

With AI, you can set up ongoing monitoring that triggers alerts when something important happens. If you follow semiconductor equipment makers, you could instruct:

"Monitor press releases, SEC filings, and Nikkei Asia news for ASML, Applied Materials, and Tokyo Electron. Alert me if any announce a capex change of more than $500M or a delay in EUV system delivery."

You could do the same for regulatory developments:

"Track FDA approvals for oncology drugs and list any public companies with products in the approved category within 24 hours of announcement."

Or for supply chain changes:

"Monitor global shipping news for Maersk and Evergreen. Alert if either reports disruptions or route changes involving the Panama Canal."

The point is you move from periodic research to continuous, event-driven awareness.

MAKING ANALYSTS WORK TOGETHER BETTER

AI can also make research collaboration smoother. Imagine a shared AI workspace where every analyst's notes, sources, and conclusions are linked back to the original documents automatically. No more emailing giant PDFs with "see page 147" in the subject line.

You can have AI auto-generate source-linked footnotes so that when someone quotes a number, it is always clear where it came from. You can maintain a "living research file" that updates as new filings or news arrive, so no one is working with stale data.

If your team covers a sector, AI can merge their separate workstreams into a unified sector brief overnight. You wake up to an integrated view rather than six siloed reports.

MANAGING THE RISKS

As with anything, there are risks to using AI in research. First, you need to avoid leaking proprietary information. If your research notes contain non-public insights or data purchased under license, you do not want that going into a public AI model.

Second, AI has limits in parsing complex legal or accounting nuances. It might miss the implications of a new lease accounting rule buried in an appendix, or misinterpret legal phrasing. That is why you set human review checkpoints before acting on conclusions.

Finally, AI is only as good as the data you feed it. Garbage in will still lead to garbage out, even if the garbage is formatted nicely.

WORKFLOW EXAMPLES FOR RESEARCH ACCELERATION

Here are a few practical prompt examples for speeding up research:

- **Example 1 (Filing Summarization):** "Summarize Procter & Gamble's 2024 10-K. Focus on new product launches, changes in geographic revenue mix, and any new risk disclosures."

- **Example 2 (Cross-File Comparison):** "Compare risk factor language between Lockheed Martin's last three 10-Ks and highlight changes related to supply chain security."

- **Example 3 (Multi-Source Synthesis):** "From SEC filings, Reuters news, and analyst reports over the last six months, summarize the investment case for Constellation Energy, noting catalysts, risks, and valuation metrics."

- **Example 4 (Structured Extraction):** "From the last four 10-Qs of Microsoft, extract Azure revenue, YoY growth, and operating margin. Present in a table."

- **Example 5 (Event Monitoring):** "Monitor patent filings for companies in the lithium extraction sector and alert when a new process patent is granted."

PARTING TIPS FOR AI-ENHANCED RESEARCH

- Use AI to point you toward the signal, not to replace your judgment.

- Keep prompts specific so the AI knows exactly what to extract.

- Always link outputs back to original sources for verification.

- Integrate multiple types of data to get a fuller picture.

- Build recurring monitoring where possible to save time.

Traditional research is like trying to read every book in the library before making a decision. AI research is like having a librarian who has already read everything, remembers it all, and can hand you just the chapters you need. You still have to decide if the story is worth investing in, but at least you are not lost in the stacks.

CHAPTER 7
ANALYSIS & VALUATION (TURNING NUMBERS INTO OPINIONS)

IF YOU WORK IN INVESTING LONG ENOUGH, you learn that for all the talk about "strategy" and "conviction," the core of what many analysts, especially the junior ones, actually do is build, maintain, and constantly tweak financial models.

The models live in Excel or some specialized software. They are color coded. They have tabs with names like "Drivers" and "Outputs" and "SensIn valuation models (like DCFs or LBOs), a "Sens tab" is shorthand for the sensitivity analysis tab. ." They contain more VLOOKUPs and circular references than any human should be forced to maintain. And they are, allegedly, the reason someone decided to pay you money.

So when we talk about Analysis & Valuation, we are talking about the beating heart of fundamental investing. You gather the data, you make forecasts, you apply valuation methods, you run scenarios, and you present the result with the confidence of someone who totally believes their 2032 EBITDA margin estimate is correct.

Generative AI is not here to eliminate this process. But it is here to make it a lot faster, a lot cleaner, and in the right hands, a lot smarter.

Let us break this down using the way this job actually works.

THE TRADITIONAL MODELING WORKFLOW

The old school modeling process in many traditional funds looks something like this:

- **Gathering historical financial data and KPIs:** Pull three to ten years of income statement, balance sheet, and cash flow data. Layer in key operating metrics like store counts, same-store sales, churn, DAUsDaily Active Users, or whatever matters in that industry.

- **Building forecasts:** Project revenue, margins, capex, and working capital. Then argue with your PM about why your growth assumption is "conservative," even though it's higher than consensus.

- **Applying valuation methods:** Discounted cash flow (DCF) if you want to feel sophisticated. Multiples and comps if you want to feel like you're honoring "market reality." Precedent transactions if you want to feel like an investment banker.

- **Testing scenarios and sensitivities manually:** Create base, bull, and bear cases. Adjust drivers one by one and watch the valuation swing around like a drunk sailor.

This is where AI comes in because a shocking amount of this is just repetitive data work.

"My kid could have done that with AI."

AI-ASSISTED HISTORICAL DATA COMPILATION

Pulling and cleaning historical data is the kind of work that makes junior analysts reconsider law school. It is repetitive, error prone, and not the place where you add real value. AI can step in and handle much of the drudgery. It can extract clean financial statements from SEC filings, earnings press releases, or those dreaded PDFs that refuse to let you copy a single number. It can standardize formats so that your P&L looks the same for every company, even if one calls it "Operating Income" and another calls it "Income from Operations." It can auto populate KPIs by scraping MD&A sections or investor decks. And when you are missing a data point, such as unit sales for 2018, it can suggest proxies, using industry growth rates and company market share data to back into a reasonable estimate. It's not that AI makes historical data glamorous, but that it makes it faster, cleaner, and far less soul-crushing.

> **Prompt Tip:** "Extract Nvidia's income statement, balance sheet, and cash flow statement for FY2020–FY2024 from SEC 10-K filings. Output as a CSV with standardized column names such as Revenue, Gross Profit, Operating Income, Net Income, EPS and USD millions. Include a separate tab for key operating metrics from the MD&A section."

Yes, you can literally ask it for CSV or Excel output. If the AI has tool access, or you download from a source like AlphaSense, BamSEC, or Bloomberg, you can skip most of the grunt work.

FORECASTING & SCENARIO ANALYSIS WITH AI

Forecasting is where judgment meets math. The math is straightforward, but the judgment is what sparks endless arguments among analysts. AI can take some of the friction out of the process by handling the mechanics. It can build baseline forecasts from historical growth rates, margin patterns, and seasonality. It can then layer in bull and bear cases automatically, flexing the key drivers up or down based on the rules you set. It can even run scenarios that combine macro and micro pressures, such as what happens to free cash flow if the Fed funds rate hits 7 percent and input costs rise 15 percent, and then stress test the results against past cycles or external shocks. The debates will not disappear, but at least the math will stop being the bottleneck.

> **Prompt Tip:** "Using Nvidia's historical revenue and margin data, create a 5 year forecast under three cases: Bull (10% CAGR revenue, margin expansion of 200 bps), Base (5% CAGR, margins flat), Bear (0% CAGR, 300 bps margin contraction).

Output in CSV with separate tabs for each case and a summary tab consolidating results."

When it comes to forecasting techniques, AI has a whole bag of tricks. It can run a simple linear regression if all you need is a trendline, or spin up something fancier like ARIMA or Prophet for time series work. If you want to see a range of outcomes instead of a single number, it can run Monte Carlo simulations and give you the probability distribution. And if you prefer something grounded in business reality, it can build driver-based models where revenue comes from volume times price rather than from vibes.

ACCELERATING DCF & MULTIPLE-BASED VALUATIONS

The discounted cash flow is the finance equivalent of flossing. Everyone knows they should do it, everyone says they do it, and at least half the time they are making it up. Multiples are quicker, more convenient, and usually close enough, which is why they get used so often.

AI does not change the fact that DCFs are tedious, but it does take away the excuse. Once you provide the assumptions, it can run the calculations instantly and hand you the output without all the Excel gymnastics. It can also suggest peer groups by scraping company descriptions and filtering for business model, geography, and size, which saves you the ritual of copy-pasting from a comp sheet you inherited from someone else.

Sensitivity analysis, the part where you see how fragile your model is, also becomes easier. Instead of writing formulas for WACC and growth rate tables, you can ask AI to generate them on the spot. The same goes for waterfall charts that show exactly how much each driver—growth, margin, capital intensity—moves the valuation.

What this really means is less time wrestling with the mechanics and more time debating the assumptions. The hard part has always been deciding what goes in the model, not getting the math to work, and AI tilts the balance back toward that harder and more useful problem.

> **Prompt Tip:** "Using the Base Case forecast for Nvidia, run a 10 year DCF with WACC of 8%, terminal growth of 3%. Output the DCF table in CSV and include a 2D sensitivity table for terminal growth from 1% to 5% and WACC from 6% to 10%."

Now, a quick detour into reverse DCF. A regular DCF asks, "Given my assumptions, what is this company worth?" A reverse DCF flips the question: "Given the current share price, what assumptions must be true?" It is a surprisingly humbling exercise, because the answer is often "revenue growth and margins that seem, at best, optimistic."

AI can make this exercise painless. Instead of manually tinkering with growth rates until the model lines up with the stock price, you can ask the system to solve directly for the implied assumptions. How much revenue growth does the market expect, what margin profile is baked into today's valuation, and how do those numbers compare with reality?

The point is not that the reverse DCF gives you certainty. It gives you perspective. If the market price assumes five years of flawless execution in a cyclical industry, that is useful to know. It tells you whether you are buying into a reasonable story or into wishful thinking dressed up in a spreadsheet.

> **Prompt Tip:** "Perform a reverse DCF on Nvidia using current market cap, net debt, WACC of 8%, terminal growth of 3%. Output the required CAGR in revenue and EBIT margins for the next 5 years to justify current EV."

NARRATIVE & QUANTITATIVE LINKAGE

Sometimes the numbers in your model and the words coming out of management's mouth do not quite line up. Your spreadsheet is happily projecting margin expansion of 300 basis points over the next three years, while management is warning about "headwinds in cost structure." Something is clearly off.

AI can help by stitching those two worlds together. It can scan earnings calls, MD&A sections, and industry research to see whether the assumptions in your model match the qualitative story being told. If your model says revenue growth will accelerate while every analyst on the call is grilling the CFO about slowing demand, that is a signal.

It can also flag the more subtle mismatches, the ones between the rosy story in the investor deck and the less flattering math buried in the P&L. That disconnect is often where the most useful questions live.

The real promise is in producing integrated outputs where the numbers and the narrative flow together. Instead of a model that lives in one tab and a strategy memo that lives in another, you get a single view of the company that links the story with the math. When those two align, you can feel more confident. When they do not, you have found the spot that deserves more of your attention.

IDENTIFYING VALUE DRIVERS & RISKS

Valuation models look scientific, with neat rows of numbers and discount rates carried out to two decimal places, but at their core they are just a bundle of assumptions pretending to be certain. The real question is which assumptions actually matter. AI can help you sort that out.

One way is through a tornado chart, which shows you which variables create the biggest swings in value. If changing the discount rate by 50 basis points barely moves the needle but a slight shift in operating margin sends the valuation flying, you know where to focus.

AI can also be useful in surfacing risks that often get buried in the footnotes. Things like supplier concentration, a heavy reliance on one geography for revenue, or exposure to a regulator who is suddenly paying attention. These are the sorts of risks that do not always make it into the neat boxes of a spreadsheet, but they can dominate the investment outcome.

Finally, AI is good at the what-if game. You can ask it to build quick modules that let you change a driver and instantly see how the impact ripples across the model. What if growth slows by half, what if margins compress, what if the cost of capital rises. Instead of rewriting formulas every time, you can explore the alternate worlds with a few inputs.

The point is not that AI makes the model certain. It is that it shows you where the uncertainty lives, which is the closest thing investors ever get to clarity.

GUARDRAILS FOR AI IN MODELING

AI is not a magic valuation machine; it's more like an intern who works very fast but occasionally makes things up. That means you need some guardrails.

First, **transparency**. If you don't know where a number came from, you can't really use it. Was that EBITDA margin pulled from a 10-K, scraped from some blog, or hallucinated by the model? You want a paper trail. So, always check the source of each key number.

Second, **auditability**. Someday, someone will ask you why your model assumed 8 percent revenue growth, and "because the AI said so" is not an answer that inspires confidence. Save the prompts, save the logic, save the inputs.

Third, **verification**. Compare the AI's numbers against something boring and reliable, like Bloomberg, FactSet, or your firm's internal database. If the machine insists Apple's gross margin is 12 percent, that's probably a clue it's looking at the wrong Apple.

Finally, **context**. AI is really good at spotting patterns in history, which is great until the industry you're modeling hits a disruption, or turns out to be cyclical, or both. The model will happily project the past forward forever. You, unfortunately, have to remember that oil prices don't always go up and housing doesn't always boom.

In other words: use AI to speed up the grunt work, not to outsource your judgment. The models can crunch numbers in seconds, but it's still your job to decide which ones matter.

FOOTBALL FIELDS, PEER MULTIPLES, AND OTHER FUN STUFF

One of the most reliable truths in finance is that if you put enough valuation methods on a slide, someone will believe you. This is why "football field" charts exist. They let you say, "Look, we valued this company six different ways, and they all kind of overlap, so it must be right." The fact that one method says the stock is worth $10 and another says it is worth $50 is beside the point. The overlap is the story.

AI can be quite good at this game. You can tell it to go find the peer multiples, EV/EBITDA, P/E, P/S, all the usual suspects, and it will dutifully trawl through filings or databases to produce a neat little table. This saves you the joy of manually pulling comps at 2 a.m., although you do need to double-check that the "peer" for Apple is not Granny Smith Orchards, LLC.

Then there is the chart itself. A football field is just a fancy bar chart, but bankers love it because it looks like evidence. Ask an AI to build one from your DCF, your multiples, and your precedent transactions, and you get a rainbow of valuation bands. You can even ask it to highlight which methods are complete outliers, such as when your precedent transactions imply a valuation five times higher than the DCF because people in 2021 were paying ridiculous multiples for anything with a pulse.

The fun part is interpretation. AI can tell you that a range looks strange, but it will not know that the weirdness comes from one peer that has 80 percent gross margins because it sells software while everyone else sells widgets. That part still requires a human to say, "Maybe let's not value Ford like it is Salesforce."

So yes, you can ask AI to do the grunt work, find the multiples, draw the football field, point out the outliers. But the judgment, the skeptical eyebrow raise, the question of "is this comp actually comparable," that is still you.

> **Prompt Tip:** "Using Nvidia and its AI semiconductor peers, compile EV/EBITDA, P/E, and P/S multiples from the last fiscal year. Create a football field chart comparing valuation ranges from DCF, multiples, and precedent transactions. Output the chart in PNG and the underlying data in CSV."

PARTING WORDS

AI will not replace the judgment part of modeling. It will not tell you when to fade consensus or double down. But it will get you to the point where you are actually thinking about the investment much faster.

In other words, less time moving numbers around and more time deciding if the numbers make sense. Your model does not have to be perfect. But if it takes you three weeks to build and someone else's AI builds the same thing in three hours, the only part of your model anyone will care about is the timestamp.

CHAPTER 8

THE STORY IS THE STRATEGY: BUILDING YOUR INVESTMENT NARRATIVE

THERE IS A DIRTY LITTLE SECRET ABOUT fundamental investing that everyone knows but rarely says out loud. You can have all the data, models, and diligence in the world, but if you cannot tell a good story, your idea is going nowhere.

The narrative is what gets people to lean in. It is what convinces the investment committee to put real money behind your work. It is what gets a portfolio manager to remember your pitch three weeks later when the stock is down 10% and suddenly looks more interesting.

If you are running your own capital, you might think you can skip this part. You cannot. Even then, the narrative is your own mental compass. It is the thing you return to when the market is screaming at you to change your mind. It keeps you grounded in your original reasoning and reminds you why you own the position.

The truth is that narrative and thesis development is not just a presentation skill. It is a core part of investing. And this is where generative AI can make you dangerous in the best possible way. The best investors are not just analysts of numbers but also architects of meaning. Generative AI, if used well, gives you a drafting partner that can sharpen your story until it resonates with both others and yourself.

WHY THE STORY MATTERS

A good investment thesis is like a movie pitch. You need to hook the audience in the first few sentences, then build a clear arc that explains the opportunity, why it exists, and why you are the one smart enough to act on it.

You might be telling it in a formal presentation to an investment committee with slides and appendices. You might be explaining it at a cocktail party to another value investor who asks "What is your best idea right now?" You might be summarizing it in one sentence for a PM in the elevator.

These are all thesis delivery moments. If you cannot tell the story in a compelling, concise way, the idea will die before it ever gets capital behind it.

The best narratives balance story appeal with data rigor. They make the listener feel that this is an opportunity that makes sense and is worth the risk, but they also back that feeling up with facts. Too much story without data and you sound like a promoter. Too much data without story and you sound like a spreadsheet.

WHERE AI FITS IN

Here is the beauty of using generative AI for thesis development. You already have the raw research. You have the filings, the transcripts, the industry analysis, the valuation work. AI can help you take that mass of notes and turn it into a clean, structured, persuasive story.

You can give it your research dump and say: "Organize these notes into a coherent investment thesis for XPO Logistics (XPO), structured into problem/opportunity, company fit, and catalysts."

It will draft an outline you can refine. You can even have it create multiple outlines so you can choose the one that feels strongest: "Generate three different thesis structures for LVMH based on these notes. One should emphasize brand strength, one should emphasize emerging market growth, and one should emphasize management execution." You get options to work with, not just one rigid output.

"WELL, ARTIFICIAL INTELLIGENCE TELLS EMPLOYEES LIKE YOU THAT YOUR PERFORMANCE NEEDS IMPROVEMENT."

FROM RAW RESEARCH TO COHERENT STORY

One of the most useful AI tricks is taking your scattered notes and forcing them into a logical progression. The classic three-part structure works well: start with the **Problem or Opportunity**, identifying the inefficiency, change, or condition creating a potential profit; then move to **Company Fit**, explaining why this particular company is positioned to exploit it; and finally highlight the **Catalysts**, the triggers that will make the market recognize the situation and reprice the asset.

If you have 20 pages of bullet points from calls, filings, and news, you can say: "Take these notes and create a three-part thesis: problem/opportunity, company fit, catalysts. Include relevant metrics and dates." This does two things. First, it organizes your thoughts. Second, it reveals where your thesis is thin. If the catalyst section is weak, you know what to go back and research.

ADDING PERSUASIVE MUSCLE

The difference between a good thesis and a great thesis is often in how it is framed. Generative AI can help here too. You can ask it to draft executive summaries, pitch memos, or slide outlines. For example: "Draft a one-page executive summary for my investment case in Shopify (SHOP) using these research notes. Focus on valuation, competitive position, and catalysts. Make it compelling for a growth-oriented investor." Or you can push it to find analogies and parallels that make the idea click: "Create three analogies for AMD's competitive positioning in GPUs that a non-technical investor would understand." It might compare AMD to a challenger airline with better routes and lower costs, or to a sports team with rising draft picks. The point is to make the complex relatable. You can also tailor the framing for different investor personas: "Rewrite this thesis for a conservative income-focused investor.""Now rewrite it for an activist investor looking for operational improvement opportunities."

BLENDING THE QUANTITATIVE AND THE QUALITATIVE

Strong theses weave together hard metrics and soft factors. You can talk about EBITDA margins and free cash flow yields all day, but you also need to address management quality, brand strength, and industry reputation. AI can help ensure these are integrated smoothly. You can feed it your quantitative model outputs and qualitative notes, then ask: "Integrate these financial metrics with the qualitative points about management and brand to create a balanced narrative."

It can also check for consistency between your story and your data. If you are claiming that margins will expand, AI can flag if your own model shows them shrinking. In this way, AI acts less like a spreadsheet and more like a skeptical colleague, always pushing you to reconcile the numbers with the narrative. The result is a thesis that not only holds up under scrutiny but also persuades.

PREPARING FOR PUSHBACK

A serious investor never walks into a pitch without knowing the counterarguments. AI can simulate those for you: "Create bear, base, and bull case narratives for Delta Airlines based on these notes."

Or: "Pretend you are a skeptical sell-side analyst. Write the three biggest objections to my investment case in FedEx." This is debate prep. If you can answer those objections cleanly, you are much more likely to win the room.

STYLE AND CLARITY

AI can be your style editor. If your thesis is heavy on jargon, you can ask it to rewrite for clarity: "Rewrite this thesis in plain English for a generalist PM without oversimplifying the technical points." It can also help you create multiple versions of the thesis:

- **One-liner:** "Boeing is a temporary mess with a permanent moat, trading at a discount to normalized earnings."

- **Elevator pitch (30 seconds):** "Boeing's current operational and reputational issues have driven the stock to a valuation that assumes permanent impairment. Yet its duopoly position in large commercial aircraft, combined with a record backlog and easing supply chain pressures, supports a strong multi-year recovery."

- **Full page:** The version you send ahead of a formal meeting, with supporting data. The extended version lays out the investment case with charts, financials, competitive analysis, and scenario modeling; essentially a document that can stand on its own if circulated ahead of a meeting.

BEFORE AND AFTER: TURNING RESEARCH CHAOS INTO A SHARP THESIS

Let's pretend you have just finished a deep dive on Deere & Co (DE). You have been through the filings, industry reports, and a dozen analyst notes. Your research file is a glorious mess — bullet points, half-formed thoughts, and scattered stats. At this stage, what you really have is a research junk drawer. Everything is in there, but nothing is in order:

- Ag equipment demand up due to food security concerns, climate impact on yields.

- Farmers replacing aging equipment after delaying capex during pandemic.

- Deere has subscription-based precision agriculture software — strong margins.

- 2024 guidance: mid-single-digit revenue growth, operating margin ~22%.

- Commodity prices volatile — could impact farmer willingness to invest.

- CNH Industrial and AGCO main competitors. Deere has pricing power but faces input cost inflation.

- Large installed base means recurring parts/service revenue.

- Precision agriculture adoption increasing, especially in Brazil, US Midwest.

- Valuation: ~14x forward P/E, below 5-year average.

- Backlog down slightly vs. last year but still historically high.

- New CEO since 2023 — focus on digital transformation, aftermarket services.

- If Fed cuts rates in 2025, financing cost for farmers goes down — tailwind.

- Deere dealer network strength — moat in distribution.

Now let's see what happens when AI helps turn that mess into a clear, compelling thesis.

One-Liner: Deere is a global leader in agricultural equipment positioned to capitalize on rising food security concerns, aging farm fleets, and the adoption of high-margin precision agriculture technology, trading below historical valuation multiples.

Elevator Pitch (30 seconds): Deere's entrenched market position, extensive dealer network, and growing precision agriculture platform give it durable pricing power and recurring service revenue. Near-term demand is supported by aging farm fleets and global food security concerns, while potential Fed rate cuts could boost farmer purchasing power. Shares trade at roughly 14x forward earnings, a discount to the five-year average, offering an attractive entry point ahead of sustained margin strength and technology-driven growth.

Full Page Thesis: Problem/Opportunity: Global agricultural productivity faces pressure from climate volatility and growing food demand. Farmers are increasingly seeking efficiency gains through technology adoption. After pandemic-era deferrals, replacement cycles for heavy equipment are resuming, creating a near-term demand surge.

Company Fit: Deere holds a dominant market share in North America with a robust dealer network and strong brand loyalty. Its precision agriculture platform adds a recurring, high-margin software and services layer to its equipment sales. The company's scale and installed base underpin its pricing power, even amid input cost inflation.

CATALYSTS:

- Continued adoption of precision agriculture, especially in high-growth markets like Brazil and the US Midwest.

- Potential Fed rate cuts lowering financing costs for equipment purchases in 2025.

- Aftermarket parts and service growth driving margin expansion.

- Digital transformation initiatives under the new CEO enhancing operational efficiency.

Valuation: At ~14x forward P/E, Deere trades below its five-year average despite high backlog levels and a clear growth runway. This presents an opportunity to acquire a durable industry leader at a relative discount.

Risks: Volatility in commodity prices affecting farmer income and purchasing behavior, competitive responses from CNH Industrial and AGCO, and potential delays in technology adoption in certain markets.

In less than five minutes, the messy research dump turned into three versions of the thesis that you can deploy in different contexts — quick hit in an elevator, conversation at a cocktail party, or full pitch to an investment committee.

REAL-WORLD AI WORKFLOWS

The real power of AI shows up not in abstract theory but in how it fits into an investor's daily grind. You can drop it straight into your workflow and use it to shape raw notes, sharpen arguments, and stress-test your thinking. What follows are some concrete examples of how prompt-and-refine loops work in practice.

- **Example 1 – Starting from scratch:** "Draft a one-page investment thesis for Zoetis (ZTS) focusing on catalysts, valuation, and industry positioning, using these research notes." Then refine: "Make the catalysts section more detailed, with dates and expected impact."

- **Example 2 – Tailoring for audience:** "Rewrite the thesis for an investor focused on dividend growth."

- **Example 3 – Creating multiple frames:** "Generate three alternative framings for the thesis in Deere & Co (DE): one centered on secular agriculture demand, one on technology adoption in farming, one on capital allocation discipline."

- **Example 4 – Opposing viewpoints:** "List the top five reasons this investment case could fail, based on historical precedent and competitive dynamics."

Thesis development is not just for formal memos. It is for those real-world moments where you have to sell the idea on the spot. When you have a PM in the elevator for five minutes, you need the elevator pitch version. At a cocktail party with other investors, you need a conversational one-liner that sparks interest. At an idea dinner, you need the polished, numbers-backed version. In every case, you are telling a story. And the story is not fluff; it is the way you connect your analysis to the decision to act. AI is not writing your story for you. It is helping you take all the pieces and arrange them so the story is clear, compelling, and credible.

PARTING TIPS FOR AI-ASSISTED THESIS BUILDING

At the end of the day, building a thesis with AI is not about letting the machine do the thinking for you. It is about using it as a drafting partner that helps you sharpen, stress-test, and package your own insights so they land with maximum impact. Here are some parting tips to keep in mind:

- Always start with structure. Even the best prose falls flat if it is not logically ordered.

- Test your thesis on different audiences to see how it lands.

- Let AI push you to consider alternative framings you might not think of.

- Use AI to check your facts and consistency before you present.

- Prepare the one-liner, the elevator pitch, and the full memo so you are ready for any context.

Data wins arguments, but stories win decisions. A great investment thesis is simply your data dressed for the occasion. Make sure it shows up looking sharp.

CHAPTER 9
THE WORK BEGINS AFTER YOU BUY

Buying the stock is not the finish line. It is the starting gun. You can spend weeks researching a company, building the perfect model, and crafting the most convincing investment thesis ever delivered to an investment committee. But the moment you own the shares, the world begins trying to prove you wrong. Markets change, management teams make decisions, competitors make moves, and governments pass laws that have a funny way of affecting your EBITDA assumptions.

The real job is keeping the investment case alive after it is in your portfolio. That means monitoring it. Not just casually glancing at headlines, but actively tracking the information that matters, separating signal from noise, and catching both risks and opportunities before the market forces you to react.

Think of it as owning a racehorse. You do not buy it, feed it once, and then hope for the best. You watch how it runs, track its condition, look for early signs of injury, and maybe even adjust the training plan when you see a chance to make it faster.

Generative AI can be the stable hand, veterinarian, race analyst, and gossip columnist for your portfolio all rolled into one.

WHY MONITORING MATTERS

Your thesis does not end at purchase. In fact, the biggest mistakes investors make often happen after they buy. They fall in love with the stock, ignore warning signs, or fail to recognize when new information strengthens the case for adding more. Continuous monitoring protects your downside by catching issues early. It also helps you identify when things are going better than expected so you can press your advantage.

For example, if you are long United Parcel Service (UPS) because you expect e-commerce volume to rise, you want to know immediately if Amazon announces a major shift to in-house delivery capacity. That is a thesis risk. On the other hand, if UPS announces a new long-term logistics contract with a major retailer, that could be a reason to increase your position.

"Meet your new investment counselor...Al Gorithm."

AI AS YOUR ALWAYS-ON NEWS DESK

One of the easiest wins with AI is creating curated news feeds for each portfolio company and sector. Instead of relying on a general Google News alert, you can build AI-powered monitors that pull updates from filings, press releases, local-language news, and analyst commentary.

You can tell AI: "Continuously track news for Deere & Co (DE) from English and Portuguese sources, categorize each update as earnings, M&A, litigation, regulation, or strategic initiative, and provide a daily summary with relevance scoring."

Relevance scoring is critical. Not every headline mentioning a company matters. If you own LVMH, a puff piece about a celebrity wearing Louis Vuitton is not as important as an announcement of Chinese luxury tax changes. AI can filter the constant chatter into a few priority alerts so you are not drowning in noise.

TRACKING THE NUMBERS IN REAL TIME

Beyond headlines, monitoring key financial performance indicators keeps your thesis grounded. If you modeled 10% revenue growth for Shopify and the company posts 7% growth, you want to know right away.

AI can automate KPI extraction from filings, press releases, and even conference call transcripts. For example: "Track quarterly active users, average revenue per user, and operating margin for Meta Platforms (META) and alert me when results deviate more than 5% from my model assumptions."

This is not just reactive. AI can visualize trends across multiple quarters, letting you spot inflection points early. If operating margins are improving faster than you assumed, you might be looking at a bigger earnings beat than the market expects.

KEEPING TABS ON THE COMPETITION

Sometimes the risk to your position does not come from your company's own performance but from a competitor's actions. If you own Delta Airlines, you want to know if United announces a major expansion into Delta's strongest hubs. If you are long AMD, you want early detection if Nvidia slashes prices on a competing product.

AI can track competitor moves in real time, scanning for new product launches, supply chain changes, or pricing adjustments. You could set a monitor like: "Alert me if any major US bank announces a savings rate increase above 50 basis points."

Supply chain monitoring is another underused tool. AI can map dependencies and flag disruptions. If you own a semiconductor packaging company, you might want alerts on any earthquake or political disruption in Taiwan that could impact upstream chip production.

WATCHING THE RULEBOOK

Regulation is a slow-moving train that can hit you if you are not paying attention. New laws, legal disputes, and ESG-related developments can all shift the risk profile of your holdings.

Generative AI can scan for new regulations affecting your companies. If you own a mining company, you might want alerts on environmental policy changes in the countries where it operates. For legal risk, AI can track lawsuits, patent disputes, and compliance violations. For example: "Monitor global patent filings for companies competing in solid-state battery technology and alert on any challenge to Tesla's portfolio."

ESG monitoring has grown in importance. Labor disputes, environmental incidents, and governance changes can all hit valuations. AI can summarize ESG-related updates with the same relevance scoring it uses for financial news.

BUILDING EARLY WARNING SYSTEMS

One of the most powerful uses of AI is in predictive monitoring. Instead of just reporting what happened, it can identify patterns that suggest something might happen.

This can involve predictive risk scoring based on alternative data like satellite imagery, shipping data, or credit card transactions. If a retailer's foot traffic drops sharply in multiple regions before earnings, you may have an early warning. AI can also run "what's changed" reports that compare the company's narrative over time. If the CFO suddenly stops talking about a major growth initiative on earnings calls, that is worth noting.

Another underused angle is monitoring management communications. If insider selling spikes at multiple executives in a short window, AI can flag it. This does not mean panic, but it is a data point worth weighing.

SEEING THE WHOLE PORTFOLIO

Monitoring is not just about individual positions. It is about understanding portfolio-level risk. AI can aggregate risk data across all holdings and model exposure to different factors. You can simulate what happens to your portfolio if oil prices spike, if interest rates drop 100 basis points, or if the US dollar strengthens against the euro. Generative AI can also run Python code, which means if you want to be fancy, you can have it perform factor analysis on your holdings to understand exposures to value, growth, momentum, or volatility. That is a whole other chapter, but it is worth noting here that your AI assistant can be as quantitative as you want it to be.

KEEPING HUMANS IN THE LOOP

The temptation with AI monitoring is to automate everything and act on every alert. That is a mistake. You need thresholds for when to intervene. A 2% deviation from expectations might not be worth trading on. A 15% deviation probably is. You also need to guard against overreacting to false positives. AI is powerful, but it is not infallible. You also need to comply with your firm's internal monitoring and reporting standards. AI can help gather the data, but the decision-making still rests with humans. Ultimately, you want AI as a sharp tool, not an unsupervised trader. The tool doesn't own the risk; you do. If the trade goes wrong, it's your name on the line, not the algorithm's. That's why the smartest use of AI is as an amplifier of good judgment, not a substitute for it.

PRACTICAL PROMPT EXAMPLES

The real power of prompting comes when you stop thinking of AI as a general-purpose Q&A tool and start using it as a way to codify your workflows. A well-structured prompt acts like a playbook as it tells the model exactly what signals to watch, how to classify them, and when to raise a flag.

This shifts the interaction from ad hoc curiosity to repeatable process, where the AI doesn't just answer questions but enforces discipline around the way information is gathered and interpreted. Think of it less as chatting with a bot and more as setting up monitoring systems you can run on demand. Here are a few concrete workflows you can start using today:

- **Example 1 – Company news feed with relevance scoring:** "Monitor global news and filings for Caterpillar (CAT). Categorize updates into earnings, M&A, litigation, regulation, or strategic initiatives. Assign a relevance score from 1 to 10 based on potential impact to revenue, margins, or market perception. Provide a daily digest."

- **Example 2 – KPI tracking:** "Track quarterly same-store sales for Starbucks (SBUX) from earnings releases and transcripts. Alert me if the result deviates more than 3% from my forecast."

- **Example 3 – Competitive monitoring:** "Track new route announcements from United Airlines, Delta Airlines, and American Airlines. Flag any route that overlaps with more than 25% of an existing Southwest Airlines route network."

- **Example 4 – ESG alerting:** "Scan local-language news in Indonesia for environmental incidents at nickel mining facilities operated by public companies. Provide an English summary of any report."

- **Example 5 – Early warning signal:** "Analyze management commentary from the last four quarterly calls for Salesforce (CRM) and identify topics that have been removed or downplayed compared to prior quarters."

WHAT AI CAN AND CAN'T ALERT YOU ON

It is worth pausing here to separate what you can reasonably expect from a general-purpose AI like ChatGPT, and what requires more specialized or technical tools. At the simple end, you can set up prompts that generate daily summaries, flag unusual phrases in earnings calls, translate local-language news, or track competitor press releases. These are relatively lightweight requests, and models can handle them inside a chat window without much extra infrastructure.

But there are limits. Free consumer apps are not continuously running in the background, pushing alerts to your inbox the way a Bloomberg terminal does. They can simulate monitoring if you ask them, but they will not autonomously wake up at 3:00 a.m. to warn you about a regulatory filing. For continuous surveillance, crisp Python scripts, and integration into real-time data feeds, you move out of the "general AI chat" lane and into the world of paid, finance-specific platforms. Established platforms like AlphaSense, along with upstarts such as OpenBB and Quantly, are built to plug directly into live market data and compliance-safe workflows, and that is why they carry a price tag.

The important point is not to confuse what sounds possible with what is operationally available in the tools sitting on your phone for free. When we say "AI can do this," we are often describing a capability in principle. To put it into production, especially in regulated finance, you usually need to pay for systems that are built for that purpose. Your free Grok app is not going to replicate a full-fledged monitoring desk.

PARTING WORDS

Monitoring and risk management is not glamorous, but it is where you protect your capital and find incremental alpha. Generative AI turns it from a slow, manual, reactive chore into a continuous, structured, and proactive discipline. The key is balance. Let AI watch everything, but let humans decide what matters. The real edge is not in getting more alerts, it is in having the right ones land in front of you at the right time. Buying the stock is the easier part. Keeping it is the art. AI is your lookout in the crow's nest, scanning the horizon for trouble and opportunity, but you are still the captain of the ship.

CHAPTER 10

TOOLS AND PLATFORMS
FOR FUNDAMENTAL INVESTORS

IF YOU HAVE SPENT MORE THAN SIX MINUTES on LinkedIn lately, you will have noticed that AI startups are now sprouting faster than mushrooms after a heavy rain. Some are giant portobellos you can build a meal around. Others are delicate chanterelles with a nice flavor but probably not something you'd base dinner on. And some are, frankly, the kind of weird fungi you are not entirely sure you should touch without gloves.

The investment world has not been immune to this fungal bloom. Every week, there is a new "AI for finance" platform claiming to be your all-in-one research analyst, compliance assistant, and latte art critic. Some will last. Many will not. The challenge, as a fundamental investor, is figuring out which tools actually help you find, analyze, and act on ideas faster, and which ones just burn your budget while emailing you "exciting product updates" about their new logo.

In this chapter, we will walk through a handful of general-purpose AI tools, some finance-specific platforms, and some weird hybrids in between. We will talk about what they are good at, what you can use them for, and why you should not get too emotionally attached to any of them. Along the way, you will see real prompt ideas, actual finance workflow suggestions, and a few philosophical asides about why you do not need to marry your data provider.

Before we begin, a caveat. These generic tools often run on specific underlying models, and they usually come in free and paid versions. The models, prices and subscription tiers I am about to describe will almost certainly change by the time you finish reading this sentence. Generative AI companies tweak their pricing models like airlines do seat fares, and the pace of change is absurd. If you are reading this in 2026 or later, treat the numbers here as a kind of historical relic, like looking back at when Netflix DVDs were $7.99 a month. The point is not the exact price but the relative tradeoffs, which will probably still apply even when the brand names and dollar figures don't.

"I see that you're very concerned about the rise of AI technology."

THE 10 GENERIC GENAI TOOLS WORTH KNOWIN

So you are a fundamental investor. You sit there with your 10-Ks, your Excel models, your Wall Street Journal, and your creeping suspicion that you will never catch up with all the documents you're supposed to read. And then somebody tells you: "Don't worry, generative AI will save you." Which is both true and, in classic finance fashion, also a little bit of a pitch.

The reality is that there are a handful of general-purpose GenAI systems that you will actually use. They are not magic stock-pickers. They are better thought of as extremely diligent junior analysts who work 24/7, never complain, and occasionally make things up. The trick is knowing which "junior" you want for which task, and whether you are fine with the intern version (free) or you want the MBA-with-a-signing-bonus version (paid).

ChatGPT (OpenAI). This is the one everyone knows, and for good reason. The free version runs on GPT-3.5, which is fast and cheerful but not always the sharpest tool in the box. The paid version, ChatGPT Plus, costs $20/month and gives you GPT-4o, which is much better at reasoning, nuance, and not hallucinating numbers from thin air. Example: you dump in the text of an earnings call, and GPT-3.5 gives you "management was optimistic about growth." GPT-4o instead tells you "they guided to 8–10 percent revenue growth, up from prior 5–7 percent, but hedged with supply chain risks." For fundamental investing, that difference is everything. Use it to draft memos, brainstorm valuation frameworks, or explain the intricacies of deferred tax assets in plain English.

Claude (Anthropic). Claude is the quiet kid in the back of the class who has perfect notes on everything. Its advantage is that it can read ridiculously long documents in one go. Upload the entire 250-page 10-K, ask "what are the new risk factors compared to last year," and Claude will calmly answer without losing track halfway through. The free version gives you access to a smaller model, Haiku, which is fine for email-sized queries. Claude Pro, at $20/month, unlocks Opus, the heavyweight model that makes ChatGPT look short-term-oriented. For fundamental investors, Claude is the weapon of choice for filings, loan agreements, and proxy statements, i.e. the stuff you know you should read but don't actually want to.

Perplexity. Imagine ChatGPT but with footnotes, like an academic paper. Every answer comes with citations. For investing, this is gold. You can ask "what's the latest on Delaware's dominance in corporate law" or "what percent of U.S. consumer spending comes from the top decile" and Perplexity not only tells you but shows you where it found it. Free version works fine for light use. Pro, also $20/month, lets you pick which engine it uses (GPT-4, Claude, etc.) and gives more queries. Use case: you're writing a client deck and you need three fresh stats to justify your "consumer spending is concentrated" slide. Perplexity finds them with sources you can drop straight into footnotes.

Microsoft Copilot. Same models as ChatGPT, but embedded directly into Excel, PowerPoint, Word, and

Outlook. This matters. You don't want to copy/paste outputs; you want Excel itself to do the work. Paid only, $30 per user per month. In Excel, you can type "build a table showing EPS sensitivity to 5 percent revenue decline and 100bps margin compression" and it just does it. In PowerPoint, you can feed it your memo and instantly get slides for Monday's meeting. It is less a research assistant and more like a super-powered macro that understands English. Fundamental investing example: quickly building bear/bull case tables inside Excel without spending 45 minutes wrestling with cell references.

Gemini (Google). Think Copilot, but for people who live inside Google Workspace. If your firm uses Gmail, Google Docs, and Google Sheets, Gemini is the natural fit. Free version is limited, but Gemini Advanced at $20/month gives you the strong model and integrations. Practical example: you paste consensus estimates into Google Sheets and ask Gemini to chart revenue versus margin trends across peers. Or you drop in a rough draft of an investment memo into Google Docs and ask it to rewrite in "hedge-fund-letter voice." Gemini is not as well-tuned to finance as ChatGPT or Claude, but the integration with Gmail/Docs/Sheets is the killer feature.

Mistral (and Meta's LLaMA 3). These are open models. They don't come with a cute app or chat window, but you can run them yourself, on-premises, or pay API costs. Why does this matter? Confidentiality. If you're at a hedge fund or bank and you don't want client data flying off to Microsoft's servers, you run your own model. Performance is comparable to GPT-3.5, sometimes better. Use case: you fine-tune Mistral on your own research archive so it can answer "what were our assumptions on oil price elasticity in 2016." Cost depends on how you host it, but the model itself is free. Compared with ChatGPT, the tradeoff is control versus polish.

Manus. Built on Mistral, Manus provides a polished chat interface with strong reasoning and speed, priced around $10–15/month. Think of it as "budget GPT-4." For investors, it is a cost-effective daily driver if you don't want to pay for both ChatGPT Plus and Claude Pro. Use case: sanity-checking your DCF assumptions. "If WACC rises 50bps and terminal growth falls 25bps, what happens to valuation?" Manus will crank out the math fast. Not as smooth as GPT-4o, but at half the price.

Pi (by Inflection). This one is more of a "thinking buddy" than a research assistant. It is less powerful at long documents, but surprisingly good at talking you through an argument. Free, with premium options on the horizon. Fundamental investing example: you're building a thesis on a beaten-down retailer and you're not sure if the turnaround narrative hangs together. Pi will happily role-play a skeptical PM, poking holes in your story until you tighten it. Its edge versus ChatGPT is tone: more coach, less search engine.

Jasper. Big with enterprises. What Jasper does best is governance: setting templates, tone, and compliance boundaries so teams produce standardized, safe content. Business plans start at about $40/month per user. For investing, Jasper is less about idea generation and more about communication. Think drafting investor letters, fund updates, or marketing materials where you don't want a rogue AI making up forward-looking statements. Compared with ChatGPT, Jasper is safer and more controlled, though less flexible.

Notion AI. If your firm uses Notion to track research ideas, manage pipelines, or keep internal knowledge bases, Notion AI is a logical add-on. Ten dollars per user per month. It won't out-reason Claude, but it will summarize and suggest inside your existing workflow. Example: you keep diligence notes on ten potential portfolio companies in Notion. Notion AI can surface "common risks mentioned across all" or "summarize last week's updates" without you leaving the platform. Compared with standalone AIs, the value is location: it lives inside the tool your team already uses.

None of these tools will replace fundamental analysis and none will tell you which stock to buy. What they do is remove friction. ChatGPT drafts your first pass. Claude chews through the filing you don't want to read. Perplexity finds the stat you need. Copilot builds the Excel sensitivity. Gemini handles it if you're a Google shop. Mistral and Manus keep things private and cheap. Pi makes you think harder. Jasper keeps you compliant. Notion AI organizes your notes. Together, they make the job less about formatting tables at midnight and more about actually thinking about businesses. Which, if you ask Warren Buffett, is the fun part.

MAPPING THE MUSHROOM PATCH: CATEGORIES OF AI FOR INVESTING

Moving on to investing-specific tools, the first thing to realize is that there is no single "AI for investing" market. What we actually have is a giant mushroom patch. To avoid wandering around like a lost hiker in the woods, it helps to put them into buckets. As of mid-2025, the buckets look something like this:

AI Agent Platforms: Think of these as the interns you always wished you had, only with unlimited patience and no need for free pizza. Tools like **Quantly**, **Hebbia**, and **Rogo** promise to read filings, comb through earnings transcripts, and then actually remember what they just read. Instead of a pile of PDFs sitting on your desktop, you get a structured output that feels like it was written by a reasonably competent junior analyst. Others, like **Menos AI**, **BlueFlame AI**, and **Matterfact**, go one step further and try to orchestrate workflows. You can say: "Screen me five companies with rising margins, build me a one-pager on each, and then draft a comparison table," and the agent will dutifully run through the steps. In theory, it is like having an analyst who not only reads but also does the grunt work, makes a chart, and staples it to your desk. In practice, sometimes the stapler jams.

Trading Platforms: Different species of mushroom here. **Axyon.ai**, **Scalar Field**, and **Fixparser** are less about reading the 10-K and more about finding signals in the noise. Imagine a friend who cannot tell you what a company does but can tell you the stock usually goes up on Tuesdays after raining in Omaha. That is the flavor. These platforms lean toward quant funds and hybrid shops that live on pattern recognition and execution speed. They are not building narratives about moats or management quality. They are looking for micro-edges, cleaner trade data, or models that say "buy here, sell there." Useful if you are wired for trading. Less useful if you like to talk about free cash flow yields at dinner parties.

Modeling and Excel Companions: This is the mushroom patch closest to home for most fundamental investors. If you live in Excel, you do not want a shiny new platform, you want Excel with superpowers. Enter **DocuBridge**, **Chatsheet**, **TenKay**, and friends. These are essentially robot analysts whose job is to clean up the nightmare of linking sheets, rolling forward assumptions, and updating line items without breaking your circular references. You can literally type into a cell: "Update my revenue forecast based on the last four quarters of growth, run a scenario with margins 200bps higher, and update the chart on Tab 3." Suddenly your spreadsheet does the work instead of you. The dream of every associate who has ever been asked to "just make one more adjustment" at 2 a.m.

Data and Infrastructure Tools: This is the plumbing. Not glamorous but absolutely necessary. Platforms like **Fiscal.ai**, **Daloopa**, **Plux AI**, and **Structify** exist for one reason: to stop you from wasting half your life cleaning data. They take raw financials, normalize them, and spit out clean numbers that line up across companies and time periods. If you have ever spent an hour trying to figure out why one company calls it "Operating Income" while another calls it "Income from Operations," you know why this category exists. Without good plumbing, all the AI fairy dust sprinkled on top is worthless. Garbage in, garbage out, as your cranky quant friend will remind you.

Research and Analysis Tools: This is the noisiest bucket because everybody wants to be in it. Here you have **Sibli**, **Tenzing MEMO**, **Hudson Labs**, **Boosted. ai**, **Benjamin AI**, **ValuWiki**, and about forty others, all claiming they can make you a better investor by automating research. Some focus on memo drafting, some on surfacing insights from filings, others on summarizing earnings calls or predicting risks. They are like mushrooms that promise to be both delicious and medicinal. Some actually are. Others are just very pretty to look at before they wilt in your fridge. The key here is that the category is massive, the tools are diverse, and figuring out which ones are genuinely useful requires testing them in your own workflow.

So what is the point of all this categorization? It is not that one bucket is better than the others. It is that the AI-for-investing landscape is no longer just one thing. You have agents, traders, spreadsheet whisperers, data plumbers, and research companions. Each category solves a different problem. Knowing which problem you actually have is more important than memorizing every startup logo you saw on LinkedIn.

If you are drowning in data chaos, start with the plumbing. If your life is Excel, start with the modeling helpers. If you want to replace some grunt work, try an agent. If you want to shave fractions of a second off execution, look at trading. If you want more insights without reading 500 pages of transcripts, play in research and analysis.

In other words, stop looking for "the one AI tool to rule them all." That tool does not exist. What you have is a Lego set. Some pieces are useful for foundations, some for details, some for making the thing look pretty. You can mix and match. The structure comes from you. And, as with mushrooms, remember: just because it grew fast does not mean you should eat it without a little testing first.

THE HYBRID ZONE: TOOLS THAT REFUSE TO STAY IN A BOX

Not everything fits neatly into the "generic" versus "finance-specific" divide. There is a middle ground, and it is full of tools that investors are quietly leaning on more than they admit. Call them hybrids.

These hybrids come in two flavors. The first is the generic tool with a hidden superpower. These tools were not built with Wall Street in mind, but somewhere in their feature set lurks the ability to crunch numbers, automate workflows, or clean data at industrial scale. If you are willing to give them detailed instructions, they will happily work through financial tasks with the patience of a junior analyst who doesn't know they are being hazed.

Julius AI is a good representative of this category. It is not technically a finance-specific platform, but it is also not just another "ask me anything" chatbot. Think of it as a Python and spreadsheet automation wizard that happens to be willing to do investment work if you tell it how. It is the tool that takes your messy CSV export from EDGAR, normalizes the dates, runs your ratio formulas, and produces a chart that looks like you spent hours in Excel.

Where Julius shines is in structured, repeatable workflows. You can script it to pull financial statements, calculate KPIs, build charts, and save the results directly into Excel or Google Sheets. That alone replaces hours of grunt work, but you can also scale it up to more ambitious tasks, like building a three-statement model for Shopify using historical data, forecasting revenue based on consensus estimates, and running a Monte Carlo simulation for EBITDA margins. Julius will not tell you what NVIDIA's Q3 guidance means for the semiconductor cycle, but it will be the quiet, tireless operator in the background building the infrastructure of your analysis. Compared to other generic tools like ChatGPT or Claude, Julius has an edge because it was designed with automation and data integration at its core. It is less about clever conversation and more about heavy lifting.

The second flavor of hybrid is the finance-native platform that has let AI run loose inside the house. These companies started out solving a narrow problem like document search or transcript retrieval but then bolted on generative AI until it seeped into every corner of the workflow.

AlphaSense is the obvious example. They began life as a search engine for financial documents, but now look more like a roll-up fund crossed with an AI lab. Their crown jewel is a library of expert network transcripts so large it feels like insider gossip with a compliance stamp. For industry sentiment, customer checks, or management tone analysis, nothing else comes close.

What sets AlphaSense apart is the combination of breadth and AI integration. You can ask it to summarize the last five expert calls on ASML and highlight recurring risks while generating a competitive landscape section for your memo, or to analyze CEO commentary on margins across several Caterpillar earnings calls and classify the tone while noting changes in forward-looking language. That is not just search; it is building an analyst's work product on demand. AlphaSense wins against generic chatbots because it has proprietary data and because the AI is not just answering questions in a vacuum, it is working on top of a curated, finance-native content set. And compared to traditional finance tools, the AI layer makes it dynamic. Instead of down-loading PDFs and running manual screens, you ask a question and watch the answer assemble itself.

Investors need hybrids because they solve two problems at once: flexibility and relevance. Julius AI proves that you can bend a general-purpose engine into doing specialized, repeatable investment tasks. AlphaSense shows how a finance-first platform can unleash AI across its data in ways that make traditional search and retrieval obsolete.

Together, they illustrate why hybrids are likely to be where most investors spend their time. The pure generic tools are useful for brainstorming. The pure finance platforms are great for compliance comfort. But in the trenches, where analysts are building models and port-folio managers are writing memos, it is hybrids that quietly do the work.

And this may be music to the ears of incumbents like Bloomberg and Cap IQ. At the end of the day, analysts already spend most of their working hours inside these platforms. If you can make their lives just a little easier by layering in AI, whether that means summarizing transcripts, automating comps, or building charts, they have no reason to leave and start tinkering with Claude or ChatGPT on the side. That is exactly why the major players have been pouring money into AI upgrades over the past few years. They know that convenience and stickiness matter more than shiny demos. If the AI lives where the analyst already lives, the battle is half won.

FINANCE-SPECIFIC TOOLS STILL MATTER

Generic tools like ChatGPT, Claude, or Perplexity are incredibly flexible. They are good at almost anything, which also means they are not particularly good at the very specific things investors actually need. That is where finance-specific platforms quietly win. They have three big built-in advantages.

The first is **compliance**. You can often set them up in private instances or keep all the processing inside a walled garden, which avoids the nightmare of accidentally uploading material non-public information into some giant public model that is busily training itself on your data.

The second is **pre-loaded data**. Many finance tools come with licensed datasets already wired in. Instead of spending weeks trying to connect your own APIs or negotiating licenses, you just get the data in the platform. That saves you both the headache and the bill.

The third is **analyst-focused defaults**. The templates, charts, and output formats are actually designed for financial analysis. They assume you want to build a sensitivity table or a comp sheet, not a social media campaign. That alone makes them easier to use when your job is to make an investment decision rather than generate marketing copy.

If you are running a small fund, you do not have to buy the whole expensive buffet right away. Start with a couple of high-value finance tools, combine them with premium subscriptions to general-purpose AI like ChatGPT Plus or Claude Pro, and then layer those on top of the Bloomberg, FactSet, or CapIQ data you are already paying for. That mix usually gets you most of the way there, without immediately blowing up your budget.

CHATGPT FEATURES YOU PROBABLY DID NOT KNOW EXISTED: OPENAI EDITION

When you use the free generic stuff, do not just type in questions and read the answers. These tools can do a lot more than Q&A, and you are leaving half the value on the table if that is all you do. Think of them less as a smart search bar and more as a Swiss Army knife that happens to live in your browser. They come with a surprising amount of hidden functionality, and the best way to understand them is to look at what is already baked into ChatGPT.

Take the **camera** feature. Instead of manually copying numbers out of a PDF, you can literally snap a photo of a chart or table and tell it to give you the data back. That turns a useless graphic into a CSV you can drop into Excel. Imagine opening an equity research report, pointing your phone at an EPS chart, and then asking ChatGPT to export the data as date versus EPS so you can run your own forecast. What used to take twenty minutes of clicking and scrolling becomes a single request.

Or consider **voice**. On mobile, ChatGPT turns into an analyst in your pocket. You can talk to it the way you would to a junior associate. Something like: give me a two-minute update on today's Asian markets, highlight any big macro moves, tell me about notable earnings in Japan, and flag anything China-related that might move semiconductors. Instead of scrolling through ten different news apps, you get a digest that is focused, quick, and portable.

Then there is **agent mode**, which is basically the ability to chain tasks together into a little workflow. Instead of you doing all the handoffs, you can just describe the sequence once. For example: download Tesla's latest 10-Q, summarize the key metrics, compare them to analyst consensus from FactSet, and then email me the results in Markdown format. That would normally take hours and several tools. Here, it is one prompt.

Connectors are another overlooked feature. They let ChatGPT pull live data from external systems. If you have a CapIQ license, you can ask it to fetch the latest revenue, EPS, and EBITDA multiples for every U.S.-listed semiconductor company. Instead of logging in, running a screen, exporting the file, and cleaning the output, you just describe the dataset you want and let the model deliver it.

The **vision** capability works like the camera example but is designed for richer images, like slides. You can upload an investor presentation, point to a specific page, and say: extract all revenue figures by region and compare them to last year's values. In other words, instead of retyping every cell in a table buried in a PDF, you let the AI do the scraping.

Another feature worth knowing is **canvas**, which is essentially a collaborative whiteboard with AI stitched into it. You can map out the skeleton of a pitch deck, ask the model to generate summaries for each section, and then export the whole thing into Word or PowerPoint. It is a way to blend brainstorming with execution so you are not staring at a blank page.

ChatGPT also has **memory**, which means it can remember your preferences if you opt in. That sounds trivial until you realize it can recall that you like

Markdown tables or that your default valuation metric is EV/EBITDA. Instead of restating those details every time, you just build up a little shared context. It is like working with a junior analyst who knows your style and does not need to be reminded.

Finally, there is **deep research**, which is what happens when you stack all the other features together. You can ask ChatGPT to find every company in the S&P 500 with gross margin growth above 30 percent over the last three years, pull their earnings calls, summarize the key growth drivers from each, and then identify common themes. That is not just automating a screen. That is automating the first draft of a thematic research project.

Once you realize the breadth of what is possible, you start to see how these so-called generic tools can bend toward finance if you push them hard enough. If you know the kinds of tasks you normally hand off to an analyst, there is probably a feature hiding inside this "generic" tool that will happily take them on.

PARTING WORDS

The temptation in AI-for-investing is to think that the tool will make you better. It will not. What it will do is make you faster at certain things, and speed amplifies both good and bad decisions. If your process is sound, these platforms are like strapping a turbocharger to your workflow. If your process is sloppy, you are just hitting bad conclusions faster.

So experiment. Play with the mushrooms. Build workflows that actually make your life easier. And do not get precious about the tools themselves because the edge is in how you use them, not in the brand name on the login screen. At the end of the day, whether you are using a billion-dollar enterprise platform or a scrappy spreadsheet add-in, the AI is just a tool. You are the investor. And if you ever forget that, just remember: mushrooms are delicious, but you still have to do the cooking.

CHAPTER 11
WORKFLOW INTEGRATION STRATEGIES FOR THE AI-CURIOUS INVESTOR

BEFORE I BECAME A FULL-TIME INVESTOR, I spent more than 15 years as a management consultant. Which means I have a suspiciously well-developed ability to make PowerPoint slides that look like they belong in an art gallery, invent new 2x2 matrices for problems that were perfectly fine in one dimension, and talk about "change management" without anyone quite knowing what I mean.

Also, I can say the phrase "aligning stakeholders" without laughing. But one genuinely useful thing you learn in consulting is that adopting new technology, whether it is a giant ERP system, a new CRM tool, or now, generative AI, is not just about learning the features. It is about embedding it into the way you work every single day. If you ask ChatGPT one question today and another question three days later when you remember it exists, that is not an AI strategy. That is digital small talk.

For investors, the optimal value from generative AI comes when it is woven into the fund's research process from the moment you sniff an idea to the day you are writing the sell note. That is where the real leverage is, in finding the little seams in the workflow where AI can shave hours off a process, surface insights you would have missed, and let you spend more time thinking and less time copy-pasting.

The first step is figuring out where in your current process AI can do the most damage in the good sense. Map the steps you take in a typical investment cycle, screening, data gathering, reading filings, building the model, writing the thesis, monitoring developments, and ask yourself: where is the time sink? Where do you have to do something repetitive, where the output is predictable, where you could tell an intern "follow this exact set of steps and give me the results"? Those are your low-hanging fruit.

From there, brainstorm where and how AI can add value at each step. Then prioritize those use cases. Which ones will save the most time? Which ones can you implement without major compliance headaches? Which ones would give you an edge if you were the only shop using them? Rank them, then pick one or two to experiment with first.

For example: maybe you spend two hours a week manually pulling segment revenue numbers from 10-Qs into a spreadsheet. That is the kind of task where a good LLM with table extraction can take a PDF, pull out exactly the fields you care about, and drop them into CSV format ready for Excel. You do not have to rebuild your valuation model every time; you just give AI the structure and let it populate.

Or perhaps you are manually summarizing management's commentary from conference calls. AI can not only summarize the call but do it with the specific focus you want, "highlight all mentions of margin pressure, pricing strategy, and capacity expansion" in Markdown so you can paste it straight into your notes.

Once you know where the wins are, the second step is to layer AI on top of the tools you already use. If you live inside Bloomberg, FactSet, or Capital IQ, you do not have to abandon them; instead, integrate AI via APIs, browser extensions, or even just by feeding AI screenshots, exports, and raw data from those platforms. For example, Bloomberg's Excel API can pull live financials, and you can point AI at that sheet and say:

"Identify the three companies in this peer set with the most volatile gross margin over the last five years, explain the likely causes, and suggest two scenarios for how margins could stabilize or deteriorate."

And it will give you something thoughtful enough to kickstart analysis faster than digging through each company's filings manually.

If you are cautious about adoption, and you should be because compliance departments love to have opinions about new tools, start small. Pick the least risky, most efficiency-gaining task, something where an error would not cause embarrassment or a loss, and get comfortable with that. Maybe it is AI summarizing industry news for your sector coverage. Then scale up. Gather feedback from your team: what is working, what is garbage, where is it hallucinating, and where is it genuinely saving time? Treat AI adoption like you would treat onboarding a new junior analyst, with structured feedback loops.

One trick to make this work better is to create internal cheat sheets. Write down the best prompts you have found for specific tasks, like "DCF sensitivity table generation" or "peer multiple comparison." Host informal brown-bag sessions where analysts can share how they are using AI, what worked, what bombed, and what could be improved. As the team gets more sophisticated, start integrating finance-specific generative AI tools, automated workflows, and proprietary datasets. That is where you move from generic value to competitive advantage.

Now, the other part of integration is getting better at telling the AI what you actually want. In Chapter 2 we talked about prompt hygiene and my lovingly over-engineered C.R.E.A.T.I.V.E.F. framework. That applies here too. You are not just asking questions; you are giving AI a role, context, and constraints that make it produce analyst-grade outputs.

One useful trick is the iterative improvement loop. Instead of asking for the perfect answer in one shot, ask AI to "show its work" — the intermediate steps, assumptions, and sources — before finalizing. Then you can say, "Tighten the valuation comps to only include companies with market caps under $5B" or "Focus only on changes from the last filing." This turns AI into a collaborative partner instead of a black box.

Over time, you will build a library of reusable prompt templates. Maybe you have one for "Earnings Call Analysis," one for "DCF Sensitivity Table," one for

"Regulatory Risk Briefing." These templates save time and ensure consistent output quality across the team. At my fund, I have seen how sharing these templates internally turns AI from a novelty into infrastructure.

The real magic happens when you combine AI's speed with human judgment. Let the AI do the first pass on a quarterly model update, pull in the new numbers, apply your default assumptions, spit out the base case, and highlight variances from consensus. Then the analyst reviews it, challenges the drivers, and adds market context.

One of my favorite uses here is having AI stress-test our own assumptions. If our model says Company X's EBITDA margin will improve 300bps over the next two years, I will ask AI:

"Give me three plausible reasons why EBITDA margins could decline by 300bps instead, citing historical precedent or comparable companies."

This is not about trusting AI over your own work, it is about using it to generate counterarguments you might not have thought of.

If you want to get fancy, you can link AI to your specialized finance tools. Imagine pulling a Buffett "playbook" from Quantly, running it against an AlphaSense transcript search for "capital allocation," and then piping that through OpenBB to pull the relevant valuation metrics. Stitch that all together in a Zapier workflow that dumps a pre-formatted report into your RMS system. You have just taken what used to be a week of scattered work and condensed it into a button click.

And do not underestimate the value of your own internal knowledge base. Every investment memo, sector primer, or post-mortem you have ever written is an asset. If you store those in a secure, AI-accessible repository, you can ask:

"Pull up all cases in the last five years where we invested in companies with >40% gross margins but declining revenue, summarize the outcomes, and note common risk factors."

That is not replacing your thinking, it is augmenting it with perfect recall.

Finally, change management. You can have the best tools in the world, but if your team does not know how to use them, or does not trust them, you will get exactly nothing. Train people not just on how to use AI, but why. Give them permission to experiment. Assign an "AI power user" to each coverage sector so they can champion adoption. And measure the impact: are you covering more names, publishing faster, spotting risks earlier?

At the end of the day, embedding AI into your workflow is not about replacing humans, it is about letting humans focus on the parts of investing that actually require judgment, creativity, and context. The rest, the repetitive, structured, predictable stuff, is fair game for automation.

But this AI software isn't REPLACING you, it's going to be MANAGING you.

MOVING BEYOND DATA EXTRACTION

So far, most of what we have talked about is the obvious stuff. You know, the kind of things every sell-side intern wishes they had: a robot that reads faster, highlights the important footnotes, and does not complain when asked to summarize twenty pages of "Other Comprehensive Income." It is useful, it saves time, and it is a relief to hand the grunt work to a machine. But if that is all you do with generative AI, if all it does is chew through filings and news so you do not have to, then you are basically running a very expensive PDF summarization service. That is not a strategy. That is a parlor trick.

The real question is what comes next once you have the basics in place and once you have taught the AI to behave like a half-decent junior analyst. This is where the fun begins. Beyond knowledge extraction lies a set of frontiers where AI can shape how you think, how you engage, how you work as a team, and how you build a lasting edge.

The first is **decision support.** Reading filings faster is nice, but the real magic happens when the AI starts poking holes in your beautifully over-engineered model. Imagine telling it: "Take our base case for Company X, now run it under a scenario where interest rates stay at five percent for the next decade." Or: "Pretend we are back in a 2008 style credit crunch and liquidity dries up. What happens to their covenants?" You do not even need to give it the exact macro inputs. It will happily invent plausible ranges and then run them through your financials. Are the results precise? No. Are they thought-provoking? Almost always. This is how you use AI not as a clerk but as a debate partner. It is the digital equivalent of the skeptical PM in the Monday morning meeting who asks: "Okay, but what if you are wrong?" Except this PM works for free, does not drink all the

LaCroix**LaCroix Sparkling Water is a flavored seltzer brand in the colorful pastel cans and is very popular in the U.S.** , and never has to be promoted.

It is also surprisingly good at reminding you of history. Humans have short memories. Analysts especially so, because there is always a shiny new stock to pitch. AI, on the other hand, has no problem dredging up examples of companies that blew up under suspiciously similar circumstances ten years ago. Tell it: "Show me comparable cases where EBITDA margins expanded 300 basis points before collapsing," and it will give you a depressing list of corporate hubris to humble your model. Which is another way of saying AI does not just save you time. It makes you slightly less likely to embarrass yourself in front of your ICIC usually refers to the investment committee, the group responsible for reviewing, approving, and overseeing investment decisions.. That alone is worth the subscription fee.

The second frontier is **campaign and engagement tools.** Fundamental investors like to pretend that once they have The Model, their work is done. But the truth is that investing is as much about persuasion as it is about analysis. You need to persuade your PM, your IC, your LPsLP is shorthand for Limited Partner. These are the investors who commit capital to a fund. , and sometimes even management. Persuasion takes work: letters, decks, memos, carefully worded scripts for that call with the CEO where you try to suggest capital discipline without sounding like Carl Icahn. This is the part nobody likes to admit, but AI is extremely handy here.

You can literally tell it: "Act as a defensive CEO and write your rebuttal to my activist letter." It will generate something close enough to what you are going to hear on the call that you can preemptively draft your counter. Or: "Translate this thesis into a two-page LP update that makes us sound like disciplined capital allocators instead of adrenaline junkies chasing turnarounds." It will do that in your house style, complete with sober phrasing. The point is not that AI will win the proxy fight for you. The point is that it can play every role in the proxy fight rehearsal room, so by the time you walk on stage you have already sparred against every conceivable counterargument. That is real leverage.

The third frontier is **workflow orchestration.** Right now, you probably treat AI like a tool you occasionally remember to use. That is fine. But the real power comes when you stitch it invisibly into the plumbing of your process. Picture this: a 10-K drops into EDGAR. Immediately, an automation fires. The AI redlines the changes from last year, summarizes the highlights, pulls the revenue segment data into a CSV, updates your model, posts the top three changes into your coverage

Slack channel, and drops a draft note into your RMS system. No human lifted a finger. By the time you finish your coffee, the grunt work is done.

You can extend this logic indefinitely. Bloomberg API spits out updated estimates, AI checks them against your model, flags discrepancies, and drafts a variance analysis table. LinkedIn shows a spike in job postings for a competitor, AI graphs the trend and pastes it into your sector tracker. It is not that any single workflow is revolutionary. It is that, like compounding interest, the time savings accumulate. What used to take twenty hours now takes two. What used to require three analysts now requires one, plus a decent set of APIs. Which is why the phrase "AI will not replace you, but the investor who uses AI will" is not motivational-poster fluff. It is a fairly accurate description of resource allocation.

Of course, all of this makes compliance nervous. And to be fair, compliance has a point. If you feed everything into a black box and let it spit outputs into your models, you are basically inviting trouble. Which is why the fourth frontier is **compliance and risk management.** Instead of treating compliance as the department of "no," use AI to make compliance automatic. AI can scan memos for unverifiable claims, cross-check tickers against the restricted list, and flag phrases that look suspiciously like MNPIMNPI stands for Material

Non-Public Information. . It can even log every step it takes, with sources, so you have an audit trail ready for when the regulator asks how you came to that conclusion. Ironically, the same technology that compliance fears most can be the thing that makes compliance more robust. Think of it as shifting from compliance as bureaucracy to compliance as background process. You never notice it, but it is always there, quietly keeping you out of handcuffs.

And finally, there is **institutional memory.** Every fund tells itself it has a culture, a process, a way of thinking. But if we are honest, most of that lives in the heads of a few senior people and in dusty PowerPoint decks that have not been opened since the last intern mislabeled an axis. Then one day your star analyst leaves, and suddenly nobody remembers why you passed on that deal in 2019, or what your view on semis was in 2015. This is where AI shines.

Load every memo, every post-mortem, every model into a secure repository, and suddenly you have a junior analyst with perfect recall. You can ask: "Show me every time we invested in a company with more than forty percent gross margins and declining revenue. Summarize the outcomes and common risk factors." Or: "What has our house view been on capital allocation in the chemical sector over the last decade?"

Instead of flipping through old files or pestering your retired partner, you just ask the machine. And it tells you. That means your fund stops forgetting. Which might be the single most underrated competitive advantage you can build.

It is also an incredible training tool. New analysts can query the archive and instantly learn how the firm thinks. They do not just read one or two case studies, they get the entire institutional brain, searchable and on-demand. It accelerates learning in a way that sitting through Monday meetings for three years never will. Which, if you think about it, is a very elegant way to compound your human capital alongside your financial capital.

Put all of these together, decision support, campaign prep, orchestration, compliance, and memory, and you start to see a new picture. AI begins as an overenthusiastic intern who has read everything but does not know what matters. Over time, if you teach it your process, stitch it into your workflows, and give it guardrails, it becomes more like a partner. Not a partner with judgment or intuition, those are still yours, but a partner that never sleeps, never forgets, and never tires of running "what if" scenarios until you are ready to throw your laptop across the room. The point is not that AI will do the investing for you. The point is that it will clear away everything that is not investing, so you can spend more of your time on the parts that actually require judgment, creativity, and context.

And that, to me, is moving beyond data extraction. It is moving from AI as a toy or a timesaver to AI as infrastructure. Infrastructure for thinking, persuading, remembering, and protecting. Which is, if you look at it sideways, the same journey we have been on with every technology adoption in finance: Excel, Bloomberg, the internet. At first it is a gimmick. Then it is a helper. Then one day you look around and realize it is the water you swim in. The only question is whether you build those workflows now, or wait until your competitor does.

PARTING WORDS

AI is like an overenthusiastic intern who has read everything but does not know what is important. If you teach it your process, give it clear instructions, and check its work, it can become indispensable. If you treat it like a magic box, it will occasionally hand you a perfectly formatted spreadsheet full of nonsense. Again, remember: AI will not replace you. But the investor who knows how to integrate AI into their workflow will.

CHAPTER 12
HALLUCINATIONS, SECURITY RISKS, AND OTHER WAYS AI CAN RUIN YOUR DAY

THE GREAT THING ABOUT GENERATIVE AI is that it talks like it knows everything. The terrible thing about generative AI is that it talks like it knows everything. It's like that guy in your freshman-year dorm who would confidently explain how the Federal Reserve works, the physics of wormholes, and the secret menu at Taco Bell all in one breath, and then you'd find out later that 40% of what he said was completely wrong, 40% was technically true but misleading, and 20% was just weirdly specific trivia about ferrets.

This is what AI people call "hallucination," which is a nice, soothing term for "lying with great confidence." And unlike your dorm-mate, the AI doesn't blush when it's wrong, it doesn't stammer, it doesn't hedge. It gives you a crisp, perfectly formatted bullet point about a revenue number that is, unfortunately, entirely made up. In investing, that's not just a harmless party trick, it's a potential career-limiting event. Imagine putting a buy recommendation in front of your investment committee based on an AI-summarized earnings call that never actually happened. And then imagine doing that in a quarter when performance is already bad.

"I RUN A SMALL INVESTMENT FIRM. UNFORTUNATELY, IT USED TO BE A LARGE INVESTMENT FIRM."

THE MENU OF RISKS

If you want to be precise, you can sort the risks of using generative AI in investing into a tidy little taxonomy. At the top are **accuracy risks**: hallucinations, mis-summarizations, numbers that sound precise but come from nowhere. Next come **security risks**: leaking confidential models, feeding a chatbot your draft activist letter, or accidentally teaching the internet your proprietary comp structure. Then there are **dependency risks**: outsourcing too much judgment to a machine that doesn't actually understand judgment. After that you get **bias risks**: subtle nudges toward certain sectors or geographies because that is what the training data contained. Add in **legal risks**: copyright, disclosure, compliance. And finally there are **reputation risks**, which matter even if every other box checks out, because investors are human and humans care about appearances. None of these risks are new in finance: bad data, loose lips, lazy analysis, structural bias, regulatory slip-ups, reputational blowups. They have all happened before. The difference with AI is that it bundles them all together, puts them on steroids, and delivers them with a smiley interface that makes you want to believe.

Hallucinations happen because large language models aren't databases. They don't "know" things in the way Bloomberg knows the closing price of a stock. They're statistical parrots that generate the most probable next word given your prompt and their training data. If you ask them a question about something that happened yesterday, and they don't have live data, they will cheerfully fill in the gap by inventing it. Sometimes they get lucky and the guess is close. Other times, they're way off.

The mitigation strategy here is boring but non-negotiable: you must verify numbers against primary sources every time. Even if the output looks polished. Even if the footnotes sound official. Even if you've run the same prompt 12 times before and it's been correct every time. This is especially true if you're using AI to do quick-turn work such as summarizing news, extracting KPIs, or generating comps. Think of it as having an enthusiastic intern who's great at formatting but not yet great at knowing when they're wrong.

Of course, hallucinations are just the flashy problem. The subtler one, and in some ways more dangerous, is security. Finance is full of things you don't want leaking: proprietary models, client data, your brilliant new short thesis that you're convinced will be the next Big Short. When you feed that into a public AI tool, you have to remember that public means public. The fact that it's sitting in a chat window with your name at the top doesn't mean it's yours forever. Many firms have rules from the SEC, GDPR, and internal compliance teams about what you can upload to a third-party service. Ignoring those is not just a "slap on the wrist" problem, it can be a multi-million-dollar, call-the-lawyers problem.

The good news is there are workarounds. Private or on-premise AI instances are becoming a thing, essentially the model lives inside your firewall and never sends your data out to the public cloud. You can also use retrieval-augmented generation setups that keep sensitive data in a locked database and only send the relevant snippets to the model at query time. Access controls, encryption, and audit trails, the boring stuff that IT and compliance people care about, suddenly become your friends.

But even with those protections, you can still run into the problem of over-reliance. AI is fast, it's smooth, it's always available, and it never sighs when you ask it to re-run a model for the third time. Which means you might be tempted to let it do everything. That's when you start to lose the muscle memory of thinking like an investor. You stop noticing that the company's "adjusted" EBITDA is actually excluding most of the expenses that matter. You forget to question whether management's five-year growth plan is plausible. You let the machine's authority wash over you. And then, one day, you realize your investment memo reads like it was written by the marketing team of the company you're supposed to be analyzing.

That's not to say AI can't be part of your process, it absolutely should be. But it should be an augmentation of your skills, not a replacement for them. Use it for the first draft, not the final word. Use it to stress-test your conclusions by asking it to play devil's advocate. Use it to scan 200 pages of earnings calls and tell you where management suddenly started talking a lot about "strategic alternatives." But keep your own judgment sharp, because that's the actual edge.

Bias is another risk worth talking about. These models are trained on huge swaths of internet and corporate data, which means they can inherit all the skew and blind spots in that data. If the model's training material overrepresents U.S. tech companies, it might subtly steer you toward U.S. tech companies as "the best ideas" even when you're looking for value in Indonesian manufacturing. If you're using AI for screens or shortlists, you have to sanity-check for systematic patterns and then decide whether you're comfortable with those patterns or whether they're leading you astray.

And then there's the legal angle. AI-generated output can inadvertently include copyrighted or licensed content, which means you might be republishing someone else's work without realizing it. More seriously, if you make a claim in a research note that turns out to be unverified AI speculation, and that note goes to clients or regulators, you could be in hot water. Compliance teams should be looped in early in the AI adoption process, not brought in later to clean up the mess.

Finally, the reputational piece. If your LPs find out that your latest blockbuster idea came from a chatbot, how will they feel about it? Some will be fine, as long as the results are there. Others may worry that you're replacing rigorous analysis with shiny toys. You can frame it however you want, "AI-accelerated research," "human-in-the-loop analysis," but the perception risk is real. Which is why transparency, in the right dose, matters.

Now, if you're reading this and thinking, "Okay, so I can't trust it, I can't tell it anything sensitive, I can't rely on it too much, it's biased, it might get me sued, and it might make my clients nervous, why am I even using it?" Well, because the upside is huge. You just have to use it with guardrails. It's the same as margin: incredibly powerful, incredibly dangerous, and most effective when you understand the risks going in.

If you want to avoid the hallucination trap, get in the habit of verifying outputs. Make "trust, but verify" your AI motto. Build workflows that automatically cross-check AI-generated numbers against Bloomberg or EDGAR before they go into your models. For security, don't just hope the terms of service are fine, work with your IT team to set up private instances or approved vendor lists. And for skill erosion, keep a healthy mix of AI work and human-only work in your week, so you're still exercising those fundamental muscles.

As for bias, you can actively prompt around it, "Give me examples from multiple regions, including emerging markets," and then cross-check. For legal and compliance risk, save your prompts and outputs so you have an audit trail if someone asks where a number came from. And for reputational risk, decide now how you want to talk about AI with your stakeholders before they decide for you.

The point here is not to scare you off AI, it's to make you respect it. This is not a toy. It's not Clippy 2.0. It's a powerful tool that can amplify your skills, but it can also amplify your mistakes. Use it like you'd use leverage: thoughtfully, with limits, and never when you're tired.

LOOKING CLOSER AT THE RISKS

Take hallucinations. It is not just the made-up revenue number, it is the way the model phrases it in precisely the tone you would expect from a sell-side analyst, complete with fake footnotes that look real enough to survive a skim. The mitigation is tedious but effective: build automated cross-checks into your workflow so you never copy-paste without validation. For security, think about the hedge fund that uploaded a custom options model into a free chatbot "just to test it." A week later, an eerily similar structure showed up in an unrelated online discussion. Coincidence? Maybe. Comforting? Not at all. The mitigation is to push for private instances, retrieval setups, and firmwide rules that treat prompts as sensitive data.

On dependency, the danger is more subtle: once your junior analyst stops building a model from scratch, they also stop understanding what drives it. Five years later, you do not have analysts, you have prompters. Mitigation here means forcing human-only reps: quarterly memo-writing without AI, investment committee debates where ChatGPT is not allowed, deliberate practice in old-school analysis. Bias? Ask any quant who has seen a factor backtest that "accidentally" favored U.S. small caps because the dataset underrepresented Asia. The same applies here. You can prompt for diversity of examples, but ultimately you must sample and sanity-check the outputs yourself.

Legal and compliance risk is maybe the driest, but also the scariest: the model that plagiarizes one paragraph of Goldman research into your client note is not just an academic mistake, it is a lawsuit waiting to happen. The fix is boring, keep audit trails, involve compliance early, and treat every AI-assisted line as if it could end up in discovery. And reputation is the quiet one. No one wants to be the fund whose big thesis was "co-authored by ChatGPT." The mitigation is not to hide it, it is to frame it correctly: AI accelerates grunt work, humans still drive judgment. Transparency on your own terms beats embarrassment on someone else's.

"I designed it with A.I."

PARTING WORDS

The fastest way to ruin your day is to take an AI output at face value without checking it. The second fastest is to upload your proprietary model into a free chatbot because "it's just testing." Both are avoidable. And if you avoid them, you get to keep the upside, speed, coverage, insight, without the career-ending downside. Fine: Trust AI the way you'd trust an intern who speaks in perfect grammar but occasionally makes up the numbers. Useful, impressive, but definitely not someone you'd let run the portfolio alone.

CHAPTER 13
GENERATIVE AI ADOPTION: FOUR WAYS IN, ONE WAY OUT

Generative AI is a bit like sex in high school. Everyone talks about it. No one knows exactly what it is. Most people aren't doing it. And those who are, are probably doing it badly. And yet, here we are. Every conference panel is about GenAI. Every board deck has a GenAI slide. Every fund manager has, at some point in the last six months, typed something like "summarize this earnings call" into ChatGPT.

This confusion is understandable. GenAI is not one thing. It is a collection of tools, techniques, workflows, hacks, hallucinations, and hype. And like most things in finance, if you don't have a framework for understanding what's actually happening, you're likely to spend your time chasing the shiny object of the month. We need a way to think about where GenAI fits into investment management workflows and how to adopt it. That way is a 2x2 matrix. Obviously.

This particular 2x2 cuts GenAI applications across two axes: the type of data you use (open or proprietary) and the nature of your process (ad hoc or structured). From that grid, four archetypes emerge. Each has a role to play. Each represents a different level of ambition and integration. Let's meet them.

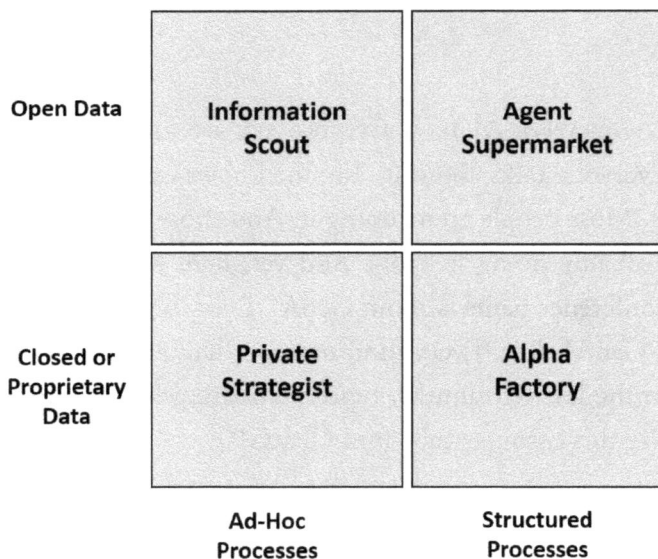

	Ad-Hoc Processes	Structured Processes
Open Data	Information Scout	Agent Supermarket
Closed or Proprietary Data	Private Strategist	Alpha Factory

THE INFORMATION SCOUT
(OPEN DATA + AD HOC PROCESSES)

This is the GenAI experience everyone stumbles into first. You have a question, you paste it into ChatGPT, and you get a glorified Google answer with better grammar. You are trying to prep for a management meeting, or make sense of a sudden stock drop, or look clever in an investment committee memo. The Information Scout is your friend.

This quadrant is full of tools like Perplexity, Claude, and Gemini. They summarize transcripts. They pull themes from earnings calls. They write your first draft of a memo that will eventually get rewritten anyway. For solo investors, this is leverage. For big firms, it's time saved. It's the AI equivalent of putting on a fresh shirt for your Zoom call. Minimal effort, maximal surface-level improvement.

But there are limits. The outputs are generic. The data is available to everyone. There are compliance risks, especially if you start uploading internal notes to public models. And there's always the lurking horror that you are unintentionally fine-tuning OpenAI with your hard-won insights about some mid-cap semiconductor firm.

Still, it's a great starting point. You learn what GenAI can and can't do. You figure out where the hallucinations live.

You start asking better questions. The Information Scout is not revolutionary, but it is how the revolution begins.

THE AGENT SUPERMARKET
(OPEN DATA + STRUCTURED PROCESSES)

Now things start getting interesting. The Agent Supermarket is where GenAI stops being a toy and starts being infrastructure. You're no longer just pasting prompts into a chat window. You're building agents. You're designing workflows. You're creating systems.

Imagine a bot that scans every same-store-sales figure in retail press releases and pushes anomalies into your dashboard. Or a script that parses pipeline drug updates across dozens of biotechs and flags major changes. This is about doing the same thing many times with consistent logic and no analyst burnout.

This quadrant is operationalized. You need guardrails. You need monitoring. You need version control. And you need to care about the output quality, because it's going straight into your research pipeline. You're not playing with fire. You're building an AI-powered toaster oven that never turns off.

The big upside here is scale. Once built, these systems run in the background, freeing up analyst time and uncovering signals that would otherwise get lost in the noise. The downside is that building agents requires technical competence, organizational alignment, and a healthy respect for the hallucination problem. You can't just duct-tape this together.

THE PRIVATE STRATEGIST
(PROPRIETARY DATA + AD HOC PROCESSES)

This is where your edge starts to show. You are no longer just another analyst feeding public data into a public model. You are integrating your own data. Expert call transcripts. Investment memos. CRM logs. Historical scorecards. The weird notes your PM jots down during meetings. All of it becomes searchable, synthesizable, and (hopefully) useful.

The Private Strategist treats GenAI not as a research assistant but as a partner. A fund uploads a decade of investment committee notes, then uses GenAI to reverse engineer what mental models led to good outcomes. A team embeds expert interviews into a semantic search engine to help frame new thesis questions. A PM runs a chat on top of 1,000 historical pitch decks to see what has or hasn't worked over time.

This isn't cheap. It requires secure infrastructure. It may involve training a custom model. You need legal to sign off. You probably need an internal data team. But if done well, this is the quadrant where differentiated insight lives. Because your data is your moat.

The hard part is that your moat needs maintenance. Data governance matters. Version control matters. And your analysts need to trust the system enough to use it, but not so much that they stop thinking.

THE ALPHA FACTORY
(PROPRIETARY DATA + STRUCTURED PROCESSES)

This is the top of the mountain. Fully integrated, end-to-end, AI-native research infrastructure. The Alpha Factory is what happens when you stop asking how to use GenAI in investing and start building an investing model around GenAI.

Here, the systems aren't just answering your questions. They're finding the questions. Your agents are monitoring app usage, traffic data, point-of-sale trends, and social chatter to proactively identify ideas. They are pushing alerts. Surfacing anomalies. Suggesting short ideas. Your analysts are curating and refining, not hunting and gathering.

It's TikTok for investing. A continuous feed of high-signal research, customized to your strategy, backed by proprietary data, and constantly improving.

Of course, none of this is plug and play. It takes years. It takes capital. It takes cultural change. You need to rethink research workflows, rebuild compliance processes, and train your team to think like product managers. But for firms that get there, the payoff is not just efficiency. It's edge at scale.

This is where new alpha comes from. Not from knowing something no one else knows, but from seeing it faster, framing it better, and acting on it sooner.

"Bring me the AI generated content!"

PARTING WORDS

You don't have to build an Alpha Factory tomorrow. You probably shouldn't. But you should start somewhere.

Use the Information Scout to save time. Build an Agent Supermarket for the repetitive stuff. Integrate your proprietary data as a Private Strategist. And keep your eyes on the Alpha Factory as a north star. GenAI is not just another tool. It's a new way of thinking about information, analysis, and decision-making. It changes what's possible. But only if you change how you work.

So draw the 2x2. Plot where you are. Pick where to go next. And if you find yourself talking about GenAI like a high schooler talks about sex, that's fine too. Just make sure you eventually learn how to use it properly.

See you in the quadrant where the alpha lives.

CHAPTER 14

THE NEXT DECADE IN FINANCE: SPREADSHEETS WITH PERSONALITY

IF YOU'VE MADE IT THIS FAR IN THE BOOK without throwing it across the room, congratulations. You have survived the parts about hallucinations, security nightmares, and the general truth that generative AI is the overeager intern of the investment world. Now it's time to look ahead. What does the future of AI for fundamental investors actually look like? Will it be more like getting a Bloomberg Terminal that can gossip or more like getting a Roomba that occasionally locks you out of your own house? Probably both.

The obvious trajectory is that AI will get faster, smarter, and much better at working in real time. Right now, a lot of the magic involves past data and after-the-fact summaries. In the near future, you will be able to ask your AI assistant, in the middle of an earnings call, "What's new compared to last quarter's guidance?" and it will whisper back, "Revenue is up 6 percent, margin guidance is two points higher, and the CFO is dodging questions about capex like it's dodgeball finals." You will get an instant "what's changed" briefing before the call is even over. Imagine that level of responsiveness applied to every piece of news and filing. SEC 8-K drops? Boom, AI flags the unusual line item in other income. FT writes a piece about supply chain risks in your portfolio company's sector? Boom, AI gives you a three-sentence risk summary, a chart of exposure by geography, and a reminder that you wrote a memo on this two years ago which it has conveniently pulled up for you.

The multimodal stuff is going to get weird in a good way. Today, you can feed AI a PDF, maybe a chart, maybe a video transcript. Soon you will hand it the full earnings webcast, including audio tone, slide deck, Q&A transcript, and even the CEO's facial expressions if you are into that sort of thing, and it will synthesize all of it into a coherent view. "CEO sounded upbeat when discussing new product launch, but slowed speech and avoided eye contact when talking about debt covenants" is the kind of sentence we may see regularly.

Whether that is alpha or just noise will depend on the context, but it is going to be part of the toolkit.

Another shift may come not from the tools themselves, but from how information finds you. Right now, the dominant model is still "YouTube style," where you type in what you want and the system fetches it. Increasingly, though, we may move toward a "TikTok style" model, where the AI knows your preferences so well that it pushes insights, data points, and ideas to you before you even think to ask. The danger, of course, is the same as with TikTok: you may end up binge-consuming dopamine hits of half-baked analysis. But the potential upside is that investors get a stream of genuinely useful, highly personalized intelligence without the friction of searching. That could change research workflows as profoundly as the jump from newspapers to RSS feeds to Twitter did.

Then there are the personalized AI research agents. These are not generic chatbots with polite greetings. These are models that have been trained on your firm's entire history of memos, models, and decisions. They know you prefer to look at trailing twelve-month EBITDA margins before gross margins. They know that when you say "small cap," you mean sub–$1.5 billion market cap, not whatever the index committee says. They know you think working capital swings are underappreciated and that you have an inexplicable fondness for companies that make industrial gaskets. These AI agents will develop a kind of institutional memory that lets them anticipate what you want before you ask for it. It will be both deeply convenient and slightly creepy, like a barista who knows your coffee order and also your recurring doubts about your largest position.

"He's asking A.I. to watch over us."

On the modeling front, we will likely see AI that can auto-build a DCF or an LBO from scratch in seconds. Feed it the latest filings, some macro assumptions, and a few comparable company tickers, and it will spit out a fully linked, perfectly formatted model with sensitivity tables, scenario cases, and the kind of conditional formatting that makes junior analysts cry tears of joy.

You will still have to check the assumptions, because nothing is worse than a perfect model built on bad inputs, but the grunt work of Excel gymnastics will become almost instant. Reverse DCFs will be trivial — just ask "What growth rate is implied by today's price given my cost of capital?" and get your answer in seconds. Football field valuation charts will be a single line of prompt. And yes, it will be able to generate the output in whatever format your CIO insists on, including CSV, XLSX, Markdown, or, for some reason, a PowerPoint table with font size 8.

All of this will make high-quality research easier to produce, which sounds wonderful until you realize what it means for competitive advantage. Historically, having ten analysts who could crank out high-quality memos gave you an edge. Now, with AI, a two-person shop can cover the same number of names. The barrier to entry for producing "good-enough" research will collapse. That means speed alone will not be enough of an edge. Your advantage will come from how you integrate AI into proprietary insights, proprietary data, and proprietary networks. In other words, anyone can ask ChatGPT for a comparable company list, but not everyone can cross-check that list against their own proprietary channel checks or integrate it with insights from private conversations with suppliers and customers.

We may also see an arms race. Early adopters will get a burst of speed and efficiency. But as more firms adopt similar tools, the speed advantage erodes, and the differentiator becomes the uniqueness of the data and the creativity of the prompts. Think of it like when Bloomberg terminals first became ubiquitous. At first, just having one was an advantage. Now everyone has one, so the advantage comes from knowing how to use it better than the other person.

Hybrid strategies will proliferate. Quant funds will start weaving in deep qualitative AI analysis of conference calls. Fundamental funds will start running factor screens in parallel with their channel checks. The lines between "quant" and "fundamental" will blur further, with AI acting as the bridge. The ability to mine unusual datasets will also expand. Imagine running daily satellite imagery analysis of retailer parking lots, cross-referenced with credit card transaction data, sentiment from social media, and supplier shipment records. Today, that requires specialized data vendors and a lot of integration work. Tomorrow, your AI agent could do it in one query.

The ESG and impact investing world will get a big upgrade too. AI will track sustainability metrics in real time, flag controversies as they happen, and benchmark companies against peers without waiting for annual sustainability reports. For better or worse, this will make ESG compliance both easier to monitor and harder to fake. If a company says its emissions are down 10 percent, your AI will cross-reference that with third-party satellite data and say, "Actually, no."

Regulators will inevitably wander into this space. They may require disclosure of AI involvement in investment recommendations. They may set standards for data governance, especially for alternative datasets. They may push for industry frameworks around ethical AI use. Whether this will be a sensible process or a circus is hard to say, but it will happen. There will be headlines about AI-gone-wrong in finance, and each one will accelerate the push for more oversight.

For early adopters, the opportunities are substantial. Operational efficiency gains will be enormous. You will be able to cover more names with fewer people, reduce the lag between an event and your response, and spend more time thinking rather than just processing. Research coverage will expand without proportional hiring. Client engagement will improve when you can provide frequent, data-backed updates without weeks of prep.

The challenge will be to avoid the trap of homogeneity. If everyone uses the same AI tools in the same way, the outputs will converge. The firms that win will be the ones that layer their own thinking, their own angles, and their own secret sauce on top. AI will handle the heavy lifting, but your unique perspective will still determine whether you generate alpha or just generate more noise.

And yes, some of the stuff we are talking about here will turn out to be overhyped. Not every promised feature will be useful. Some will be like those smart fridge touchscreens that seemed futuristic but ended up mostly showing the weather. Others will quietly become indispensable. You probably will not remember a time when your AI assistant could not instantly pull up every instance in the last decade when a CEO used the phrase "strategic review" in a call.

In the end, the future of AI for investors is not about replacing people. It is about amplifying them. The analyst of the future will be part detective, part storyteller, part systems engineer, and part AI wrangler. The skills will shift, but the core of investing — judgment, insight, patience — will still belong to humans. At least until the AI starts filing its own 13Fs.

PARTING WORDS

The firms that win in this new environment will not be the ones that adopt AI the fastest, but the ones that weave it so deeply into their process that it feels invisible. Asking AI a question once a week will be like going to the gym once a month. Technically you are participating, but you are not going to see results. Integrate it into the daily flow. Train your agents. Build your proprietary knowledge base. Iterate. And always, always check the numbers. One last thing: The future is not man versus machine, it is man with machine versus man with a slightly better machine. So get the better machine.

EPILOGUE:
THE SPREADSHEET STILL TALKS

So. You made it. You survived the hallucinations, the prompts, the fake confidence of chatbots, the real uncertainty of your own profession, and hopefully at least one moment where you opened an LLM and thought, "Wait, this is actually kind of useful."

You've learned about fire. About Prometheus. About 2x2 matrices that consultants love and investors pretend to hate but secretly love too. You've met scouts, agents, strategists, and factories. You've been warned not to paste sensitive data into a chatbot and encouraged to let AI summarize your next compliance memo anyway. Progress.

But here's the real point.

This book was never about how to master generative AI. That's impossible. The thing keeps changing. What worked six months ago is already obsolete. The models get better, then worse, then better again. The prompts you thought were clever now look like training wheels. Mastery is a myth.

What's real is the mindset. The willingness to experiment without being reckless. To move fast without breaking compliance. To stay curious without getting distracted by the hype. To treat GenAI like a junior analyst who's kind of brilliant but also doesn't know how to use a spell-checker or understand context. Useful. Impressive. Not portfolio-ready on its own.

Also, let's be honest. Most investors won't bother. They'll read a few blog posts, delegate the tech stuff, and keep doing things the old way until they wake up one day and realize their edge disappeared quietly, one summarization at a time.

But not you.

Because you read this. Which means you've already done more than most. You've started building a mental map. You've considered the risks, explored the frameworks, and maybe even tried a few tools. That's the win.

Now, before we part ways, remember that the story doesn't end here. The models will change, the use cases will expand, and the edge will keep shifting. The only sustainable advantage is staying in motion. So test the tools, break a few mental models, and keep showing up curious. That's how you make sure you're still in the game when the next wave hits.

Thanks for reading. Stay sharp and don't trust the interns with your P&L. They still hallucinate.

See you out there.